Engineering for climatic change

Proceedings of the symposium on engineering in the uncertainty of climatic change, organized by the Institution of Civil Engineers and held in London on 28 October 1992

Edited by Roger White

Thomas Telford, London

Conference organized by the Institution of Civil Engineers and co-sponsored by the British Hydrological Society, the Construction Industry Research and Information Association, the Department of the Environment, the Institution of Structural Engineers and the Institution of Water and Environmental Management

Organizing Committee
R. White, Sir Alexander Gibb & Partners (Chairman); P. Andrews, Northamptonshire County Council; M. Barrett, Consulting Engineer; M. Beran, Institute of Hydrology; C. Collier, Meteorological Office; J. Gammon, Dar Al-Handasah Consultants; J. D. Williams, Consulting Engineer; W. Wilson, James Williamson & Partners

First published 1993

Distributors for Thomas Telford books are
USA: American Society of Civil Engineers, Publications Sales Department, 345 East 47th Street, New York, NY 10017-2398
Japan: Maruzen Co. Ltd, Book Department, 3–10 Nihonbashi 2-Chome, Chuo-ku, Tokyo 103
Australia: DA Books and Journals, 648 Whitehorse Road, Mitcham 3132, Victoria

Classification
Availability: Unrestricted
Content: Collected papers
Status: Refereed
User: Civil engineers

A CIP catalogue record for this book is available from the British Library.

ISBN: 0 7277 1925 4

Published on behalf of the Institution of Civil Engineers by Thomas Telford Services Ltd, Thomas Telford House, 1 Heron Quay, London E14 4JD.

Printed and bound in Great Britain by Redwood Books, Trowbridge, Wiltshire

Preface

R. L. WILSON, President of the Institution of Civil Engineers
1991–1992

The Institution of Civil Engineers, as a learned society, has for many years provided informed opinion upon a range of matters of public concern. The climate change issue bears upon all sections of the civil engineering profession, and I hope this volume deals with this issue with an appropriate breadth of vision. We are especially pleased to include the keynote paper by Sir Crispin Tickell, an international authority on the implications of climate change. The scene is set by three eminent speakers in the fields of meteorology, hydrology and oceanography, and the scenario they present provides a backcloth for the papers that follow. These papers show that the civil engineering profession is taking the question of climate change very seriously. They consider whether engineers should respond immediately or wait and see how things develop and their answers vary according to the type of engineering being considered.

Foreword

ROGER WHITE, Editor

There is a remarkable consensus in the scientific community about the warming of our climate, arising largely from the build-up of carbon dioxide and other greenhouse gases in the atmosphere.

Engineers are at one stage removed from the science of climatic change. They are concerned with the implications of this science in terms of reservoir yields, provisions for flood defences, wind loads, soil moisture, demands on energy and water supply systems, and other factors affecting the design and construction of engineering works.

The papers which follow are remarkable in presenting both a distillation of the scientific wisdom and the responses of practising engineers.

Distinguished expert comment on the scenarios of change is provided by the Director of the Meteorological Office's Hadley Centre for Climatic Prediction and Research, the Director of the Institute of Hydrology, and the Director of the Proudman Oceanographic Laboratories.

The scenarios describe the likely range of changes over the next 40 years, both globally and in the UK, for temperature, evaporation, precipitation, storminess and sea level. The paper on scenarios was circulated to all authors and it provides the basis for the engineering responses expressed throughout this volume.

The engineering responses cover a broad sweep of the profession, which is represented in water resources and water supply, energy, maritime, geotechnics and structures.

The implications of climatic change vary, as do the responses. In the water and maritime sectors it is clear that planners are giving careful attention to the potential changes and appropriate adjustments are being made to the design of works.

Close monitoring of the situation will be essential if we are to avoid being caught unawares. It will be for the individual sectors to consider further what steps they should be taking and in so doing they will find food for thought in the prudent policy of 'no regrets' which is set forth in the paper by Hewett, Harries and Fenn.

Contents

Global warming and its effects

Sir CRISPIN TICKELL, Warden, Green College, Oxford

Climate is invariable only in its variability and, of course, uncertainty. Such variability is a prime factor in the process of evolution. Usually change takes place so slowly that we do not notice it. Most animals, plants and other forms of life have time to adapt or migrate. The Thames Valley is a good example: 130 000 years ago it was the habitat of swamp-loving hippos; 18 000 years ago reindeer and mammoth roamed the tundra of Primrose Hill and only 900 years ago the French were trying to close down vineyards in southern England which were too competitive.

Variability

We are living in one of the brief warm periods in the recent history of the Earth. For the last 2.5 million years there has been a broad 100 000 year rhythm, with over 20 glacial periods interspersed with 10 000 to 15 000 year interglacials. The cycles are the product of a combination of Milankovic cycles (in which the proximity of Earth and its angle of rotation with respect to the sun vary); tectonic plate movement and the rise and fall of mountains (which affect the relationship of land and sea); changes brought about by volcanic emissions; and the influence of living creatures, for example algal blooms, which can provide the seeding for clouds. There is variation within warm periods. For example, during the hypsithermal, about 6000 years ago, the average temperature was probably about 1.5°C more than it is today. Sea levels were a little higher. Cooler conditions then set in, and did not reverse until around 2000 years ago. The early Middle Ages were the warmest in recent memory. But in the little ice age between the fourteenth and eighteenth centuries temperatures were between 1 and 2°C below those of today. There were also wide regional variations, and we still know relatively little about corresponding events in the southern hemisphere.

Within this broad picture there is clear evidence of human-induced or anthropogenic change. Even before the industrial revolution, and as far back as the Bronze Age, there were variations in microclimate due to changes in land use, in particular deforestation. There is a close relationship between trees and rainfall. Without a quantum of trees there is less

Engineering for climatic change. Thomas Telford, London, 1993

1

than a quantum of rainfall. Examples are in the eastern Mediterranean along the coast of North Africa, and in Arizona. In modern times the changes wrought by desertification and urbanization are evident on every continent except Antarctica.

The principal anthropogenic factor in climate change is, however, the greatly increased emission of greenhouse gases — carbon dioxide, methane, nitrous oxide, etc. — into the atmosphere. Without a greenhouse effect Earth's temperature would be around 33°C cooler and life as we know it would be impossible. Carbon dioxide accounts for the largest proportion of greenhouse gases. We know from the analysis of ice cores that during the last glaciation the concentration in the atmosphere was 190 ppm. It had risen to 280 ppm before the beginning of the industrial revolution. It is now 355 ppm and rising by 0.5% annually. The main source is combustion of fossil fuels and biomass, in particular forest burning.

The IPCC findings

Public anxiety about the possible effects of environmental and climatic change have greatly increased over the past 20 years. Obviously some measure of scientific consensus on the key factor of climate was essential. Hence the creation of the Intergovernmental Panel on Climate Change (IPCC) in 1987. The IPCC published its first assessment in 1990. On what it termed a 'business as usual' scenario, the Panel predicted on the basis of modelling an average rise of around 1.0°C by 2025, and 3.0°C by the end of the next century. This would not be steady because of the interplay of many other factors. Such a rise might well seem small but there would be wide regional variations, and it can be compared with a drop of only 5.0°C in the last ice age.

In such circumstances the Panel also predicted an average rate of global mean sea level rise of about 6 cm per decade over the next century (with an uncertainty range of between 3 and 10 cm per decade). This would be caused by thermal expansion of the oceans and the melting of some land ice. The rise could thus be about 20 cm by 2030, and 65 cm by the end of the next century. Not all models confirm this prediction. In the past, warmer climates have brought more precipitation, including more snow at the poles, and thus thicker ice. But even if sea levels fell as a result, the effect could be temporary as warming continued, and most models show a substantial increase as in previous warm periods in Earth's history.

There would be markedly different results in different places. Temperature change is always more pronounced over land. So the northern hemisphere varies more than the southern. In the north the most marked increases have so far been in minimum nighttime temperatures rather than maximum daytime temperatures. Southern Europe and North America would have less summer precipitation and lower soil moisture. But in general there would be more precipitation. Snow cover and ice would generally be reduced. Modelling capacity cannot predict precise geographical results but changes would be smaller but more rapid in equatorial areas than in temperate areas.

At the beginning of this year the Scientific Working Group of the IPCC met in China to check the 1990 conclusions in the light of new information and continuing warming trends worldwide. There were no startling differences, but some interesting new evidence was produced, in particular about the short-term cooling effect of sulphates and other man-made pollution in the lower atmosphere (especially in the north). Airborne particles such as those from volcanic eruptions (in the upper atmosphere) or sulphur emissions (in the lower atmosphere) can affect radiation scattered and absorbed by the atmosphere, and also affect the microphysics of clouds, increasing cloud cover and therefore increasing reflectivity. Both of these have an overall cooling effect. The sulphur emissions in the northern hemisphere may have offset a significant part of the greenhouse warming in that region. This was previously recognized but there has now been progress in quantifying the effect.

The rate of increase of atmospheric concentrations of methane from paddy fields, termitaries, coal mining, cattle, etc. has slowed somewhat although it is not clear why. The role of methane in climate change is anyway mysterious. With huge reserves of methane locked in hydrates under permafrost and beneath the ocean, it may have played a major role in switching from glacial to interglacial periods in the past. Large-scale releases of methane could be equally dramatic in the future.

Nitrous oxide has continued to increase in the atmosphere. It comes largely from biomass and nitric acid. A more potent greenhouse gas is constituted by those convenient molecules, the artificial chlorofluorocarbons and halogens. Indeed molecule for molecule each has 12 000 times more warming capacity than carbon dioxide. But as CFCs indirectly destroy ozone, another greenhouse gas, the total warming effect may be neutral.

Uncertainties

There are many uncertainties, but none, singly or together, affects the main predictions. Among the uncertainties are: variations in solar

3

radiation; clouds and the hydrological cycle; the role of the ocean as a thermostat and its exchanges with the atmosphere; the nature of the carbon cycle; and the behaviour of polar ice sheets and sea ice. Things could prove better or worse in the event. The processes of feedback, positive or negative, are not fully understood. It is easy to be optimistic but disasters, short or long term, have happened in the past. Above all, complex and only partly understood non-linear systems can spring surprises.

It is the speed of change which is important. According to the IPCC, the changes will be greater now than have occurred normally in the last 10 000 years, and the rise in sea level will be 3 to 6 times faster than in the last 100 years. We are already embarked on an unstoppable course of global warming. Even if all human-made emissions of greenhouse gases were to be halted, about half the increase in carbon dioxide concentration caused by human activities would still be evident at the end of the next century. We need more observations on concentrations of greenhouse gases, temperature, humidity, precipitation, clouds, the radiation budget, oceanic parameters and circulation and the cryosphere.

International approach

Against this background climate change and its effects became one of the most important points on the agenda of the Earth Summit at Rio in June 1992.

- the preparation and signature of the Climate Convention
- its character and defects
- attitude of the industrial countries
- attitude of the rest
- relative vulnerabilities in the future.

Principal effect of the changes

First we must recognize that climate change is only one among many closely related changes to the total global environment. The single most important one is the astonishing success of our species in reproducing itself. Like any species that does well, whether caterpillars, thistles or lemmings, we risk going beyond a sustainable threshold. At the end of the last ice age, when humans began to spread into the Americas, the population was roughly 10 million. In 1930 it was 2 billion. It is now 5.3 billion. Unless a world catastrophe occurs, it will be 8.5 billion in 2025

As a species humans already appropriate some 40% of Earth's total net photosynthetic production on land; and land deterioration is of course linked to human proliferation. The World Resources Institute in

Washington has recently produced a report which shows that approximately 10% of the vegetation-bearing surface of Earth is suffering moderate to extreme degradation due to human activity since 1945. That is an area roughly equivalent to China and India combined. Using up non-renewable resources like coal and oil is something we do every day in the hope that we shall be able to find substitutes when the time comes. But overuse of renewable resources, like forests, is worse because it may render them non-renewable.

There is a particular problem over water. Demand for fresh water is continually increasing. It doubled between 1940 and 1980, and will double again by the end of the next century. As an illustration, all states bordering the Nile plan want — indeed need — to take out more water, but the total volume has recently diminished, and there is no way in which demands can be met. It is hard to imagine the mighty Nile turning into a sickly salty trickle like the Colorado when it reaches the sea.

What sort of world do we have in mind, say by the middle of the next century? Pressure on fertile land, and thus food supplies for a still growing population will come from many quarters: shifts in climatic zones, changes in methods of crop, livestock and fish-farming, less good water for irrigation, and loss of arable land through desertification or rise of sea level. Semi-arid areas are particularly vulnerable to droughts or floods. Of course biotechnology may come to the rescue, and enable us to make better use of the land for food. Some plants flourish in an atmosphere richer in carbon dioxide, although we have yet to see results in terms of productivity. But it is hard to avoid the conclusion that food, particularly in poor countries, will be in shorter supply, and at the spiral of deprivation and poverty will be given a new twist. Those countries with ramshackle administrations, poor internal communications, and lack of human and technical skills will be particularly hard hit.

We would lose ecosystems which could not move or adapt themselves quickly enough to new climatic regimes. Island, mountain, coastal and isolated ecological communities would be at special risk. Much would depend on the fate of certain key species, from phytoplankton to parasitic wasps. Organisms with rapid reproduction rates and high adaptability would obviously do better than longer lived organisms set in their ways. So micro-organisms and insects should be more successful than elephants and trees. As the motor of evolution works faster in warm conditions (more species, fewer individuals) than in cold ones (fewer species, more individuals), the whole system could speed up in a warmer world. We must also reckon with the genetic effects of increased ultraviolet radiation through the weakened shield of the ozone layer.

The human impact: refugees, health and engineering

Not unnaturally, we have a particular interest in our own fate, the more so as it will be a product of human interference with natural processes. Looking only at the climatic aspects we would suffer from almost any disruption of existing trends if it took place as quickly as seems likely. The cards of national advantage and disadvantage would be redistributed. But as supremely adaptable and ingenious mammals, we could modify our practices over time, especially in countries stretching over several climatic areas, equipped with modern technology. In the past we have been able to respond with our feet but now we no longer have room for manoeuvre.

We would be particularly vulnerable to sea level rise. A substantial proportion of the human population lives in low-lying areas, which would be flooded or liable to high tides or storm surges. Rising sea levels would result in raised water tables, and salt content and contamination of fresh-water supplies. Changes in water levels and precipitation would not necessarily be in step with changes in temperature. We would also suffer from changes in other terrestrial ecosystems: notably from proliferation of fast adapting organisms, and from variations in food supplies from plants and other animals.

Another broad result would be a vast increase in human displacement. In 1978 there were five million refugees on political definition; in 1989 14.5 million on the same definition; now there are up to 17 million. Add some 10 million environmental refugees or economic migrants, and the current total is around 25 million. With disruption of existing patterns of life, numbers on the move could increase drastically. It is not fanciful to estimate that with world population rising beyond 8 billion, the number of refugees could accelerate at an even higher rate with alarming conse-quences for the integrity and good functioning of human society as such.

Of more direct interest are the risks to public health. Temperature and moisture are determining factors for biological agents in the human environment: in water, food, air and soil. Variations in both affect the ability of viruses, bacteria and insects to multiply and prosper. History is full of examples of societies and civilizations brought down by diseases to which local populations had no immunity. The Black Death reduced the medieval population of Europe by between a third and a half, and smallpox and measles that of the indigenous population of the Americas by over three quarters. So it would be wise to expect changes in patterns of diseases, particularly in populations already debilitated for other reasons. Thus we could see the spread of such non-parasitic diseases as yellow fever, dengue, poliomyelitis, cholera, dysentery, encephalitis, tuberculosis and pneumonia and of such parasitic diseases as malaria, leishmaniasis, schistosomiasis, hookworm, tapeworm and other helminthic afflictions. We must also reckon with increasing contamination of water supplies, including problems arising from drainage and sewage disposal, algal blooms from nitrate pollution, salinization and aluminium toxicity.

I scarcely need to draw attention to the role of the engineers.

- fresh water supplies
- management of water grids to take account of changes in rainfall
- likely requirements of river management here and elsewhere
- new irrigation systems and reconstruction of old ones
- sea level rise
- questions of policy
- sea walls, surge barriers, rubble breakwaters, etc.
- managed retreat through encouragement of salt marshes, mangrove plantations in the tropics, etc.
- energy
- new methods of power generation: sun, tide, wind, nuclear
- better non-polluting use of old methods: e.g. coal gasification and sequestration of CO_2
- geotechnics and structures
- soil shrinkage and swelling
- aquifers
- floods and hurricanes

In industrial countries we must export many new mega-projects like the Channel Tunnel, the waterway link between the Rhine and the Danube, and the planned rail system under the Alps. Clearly we need to look at the mega-projects of the past, like certain hydroelectric schemes and the Aswala Dam, which arguably have done more harm than good, and grossly neglected environmental considerations. The latest cases are the Rarmada Dam in India and the Three Gorges Project in China.

Conclusion

In conclusion I make four fundamental points:

- We are not dealing with one factor — climate — but with a vast combination of factors: population increase, environmental degradation, and loss of biodiversity
- Science, like life itself, is full of surprises — some good, some bad. Few things go in straight lines. Life prefers leaps to gradients

7

- The problem is not change but the rate of change
- Human society is fragile. All previous civilizations have collapsed. Ultimately we are as subject to biological restraints as any other animal species. But unlike them we can consciously shape our future. If we fail to do so, there will be no-one to blame but ourselves.

The likely effects of climate change in the United Kingdom and elsewhere

M. BERAN and N. ARNELL, Institute of Hydrology, Wallingford, and C. COLLIER, Meteorological Office, Bracknell

1. Background

The earlier generation of climate models have assumed a steady $2 \times CO_2$ condition in the atmosphere and so provide no information on the transient state that will occur as the earth gradually warms through the first third of the next century. However, recent models operating in "transient mode" have revealed how the climate response to continuously increasing greenhouse gases is extremely irregular with intermittent periods and significantly large regions where cooling actually replaces warming. Such irregularities will have a significant impact on the types of impact of interest to engineers, and they certainly add greatly to the difficulty of framing a scenario for future change. For the moment practical scenarios are based on the results of the earlier steady-state versions of GCMs.

A scenario is about how a system might reasonably be expected to behave under certain circumstances. The attributes of a scenario are

- It must be internally consistent
- It must not conflict with known facts
- It should be constructed in such a way that new facts as they are discovered can be incorporated
- As a planning tool it must relate to variables and processes of practical interest

In what follows we attempt to set out a scenario for the effects of climate change, primarily for the United Kingdom, but broad indications are also expressed for other regions of the globe. The climate information has been taken directly or extrapolated from the Intergovernmental Panel on Climate Change Scientific Assessment (published in 1990 by Cambridge University Press), and the report of the UK Climate Change Impacts Review Group (published in 1991 by HMSO).

2. Global and regional results of doubling the present level of atmospheric CO$_2$

Numerical models of the global climate indicate that, in equilibrium with doubled atmospheric carbon dioxide, the following broad scale changes of climate are likely:

(a) A global mean surface warming of about 3 – 4°C, and more confidently between 1.5 and 4.5°C. Most of the larger warmings are obtained by the recent models, and are likely to have resulted from the more comprehensive treatments of clouds.

(b) A stratospheric cooling of between 3 and 5°C.

(c) An increase in warming with height in the tropical troposphere.

(d) Enhanced warming at the surface in high latitudes in winter. In the models, the magnitude of this feature depends on the treatment of sea-ice and albedo, and the resulting feedbacks.

(e) Generally greater precipitation in equatorial regions and in the middle and high latitudes, and a tendency to decreased precipitation in the tropics away from the equator.

Of course several of these more reliable 'greenhouse' predictions have little or no implication for civil engineering. Moreover it is difficult to assess the likely regional distribution of climatic changes and, at present, the indicated changes are not reliable. If the global results are correct we must assume that a mean surface warming of between 1 and 4°C is probable in the UK, accompanied by increased precipitation in winter amounting to up to 10%, such a shift in the mean, even without any change in variability, may lead to increases in the frequency of some severe events and threshold exceedances. Other consequences for elsewhere in the world are discussed later. Finally it should be noted that inter-annual variations of up to 2°C in the annual-mean surface air temperature over western Europe have been observed in some decades. Hence natural variability could mask any warming due to increased greenhouse gases.

3. Scenarios for offshore and coastal engineering

Prediction of future sea level rise is complicated by the interaction of uncertainties related both to future climate change and geological factors.

The best estimate of global sea level rise is 14 – 24 cm, but it is possible that sea level could rise by as little as 4 cm by the year 2030. Because sea level is also influenced by factors such as land movements, sedimentation, subsidence due to groundwater pumping, as well as lingering crustal movements following the last era of glaciation, regional changes in sea level may differ from the global average. We must also note that even if the greenhouse gas emissions were halted, global sea level is expected to continue to rise for may decades, and possibly for hundreds of years, due to the long response times of the polar ice sheets and the slow processes of heat transfer from the atmosphere to the ocean. It would be prudent therefore to assume a 14 – 24 cm sea level rise for the UK coastline by 2030: this is at least double that historically recorded around our coast.

Mid-latitude storms are driven by the equator-to-pole temperature contrast, and since this contrast will probably be weakened in a warmer world, it has been argued that mid-latitude storms may weaken or change in their tracks. However, present models do not resolve small-scale disturbances and so it is not yet possible to assess changes in storminess. We suggest a scenario in which there will be slightly more frequent severe storms over southern England, and a slightly reduced frequency of such storms over Scotland, say a 5% increase and decrease respectively.

We may expect changes of climate to impact upon the frequency, intensity and, perhaps, location of both tropical and mid-latitude storms. Tropical storms such as typhoons and hurricanes only develop at present over seas that are warmer than about 26°C. Although the theoretical maximum intensity of such storms is expected to increase with temperature, climate models give no consistent indication whether they will increase or decrease in frequency or intensity.

4. Scenarios for civil engineering structures on land

It is proposed that the changes in Table 1 be assumed when addressing the impact of climate change on civil engineering structures in the United Kingdom over the next 40 years. Precipitation refers to either snow or rainfall. Since mean temperature is expected to rise we might expect snowfall to decrease. However, it is possible that the overall incidence of wet snow could increase slightly which might be significant for roof loading calculations.

5. Hydrological changes in the UK

Changes in average hydrological characteristics are very dependent on the assumed scenario for climate, especially rainfall; and at the monthly

Table 1

Temperature/ocean parameter	Change over next 40 years
Temperature	1-4°C
Potential evaporation	+ 10% in summer
Precipitation - average	+ 10% in winter No change in summer + 2% increase in wet snow
- extreme	increased frequency 1 in 50 years becomes 1 in 40 years (say)
Storminess (high winds)	reduced slightly (- 5%) in Scotland increased slightly (+ 5%) in S England
Sea level rise	14-24 cm
Temperature inversions at low level	probably a decrease in frequency of a few per cent

time scale are additionally influenced by the nature of the catchment, especially its geology. The case is considered of an increase in winter rainfall of 10 per cent with an intermediate pattern of change in spring and autumn. In addition we have assumed the temperature rise leads to an increase in evaporative loss of 10 per cent. A summer reduction in precipitation has not been considered but would be expected to aggravate slightly the situations described in the following paragraphs.

River flow

Over a wide range of UK catchments such a scenario would lead to an increase in total annual run-off of around 5 per cent. In one climate–geology combination, of a type found in the drier areas of lowland Britain, the seasonal pattern of change showed flows 8 per cent higher during the winter and 4 per cent lower in summer. In another catchment of similar climate but with a large chalk groundwater contribution, flows during the summer were sustained by the higher winter inputs and a 5 per cent increase in river flow was found throughout the year. In a cool upland catchment which, in the current climate experiences a close balance between summer rainfall and evaporation, the warmer summer tips the water balance into deficit with summer flows reduced by 15 per cent. Flows in the remainder of the year were 5 per cent higher. The large divergence between catchment response emphasizes the need for studies of impacts on river flows to pay attention to the individual characteristics of the study basin.

12

Groundwater and aquifer recharge

Changes in groundwater recharge are dependent on the extent to which a shorter recharge season (following a drier summer) outweighs greater winter recharge potential. An additional consideration is the prospect of altered infiltration characteristics of some soils modified by drought or by more humid winters. Little quantitative work has been done so all that can be said is that, saving these special ground situations, it is unlikely that annual groundwater recharge would decrease given a 10 per cent precipitation increase. Special cases where consequences would be negative include aquifers whose exploitation depends on sporadic summer recharge, and coastal aquifers within the reach of saline intrusion.

Soil moisture and evaporation

Potential evaporation increases by 5 to 6 per cent for each degree temperature rise. Therefore warmer summers with high evaporation lead to longer and deeper summer soil moisture deficits, with the seasonal pattern very much dependent on the details of the changes in spring and autumn rainfall.

Plant water use and carbon dioxide stimulation

Many plant species respond to increased atmospheric CO_2 which stimulates plant growth and increases plant water use efficiency. Modelling work which includes this effect tends to suggest that the reduction in water use by plants largely cancels the increase in evaporation due to warming. Despite this apparently positive impact, scenarios have conventionally not incorporated this phenomenon. This is because our knowledge is based on results for plants growing under ideal conditions and it is likely that response in the natural environment could be very different where plants acclimatize and are subject to other stresses and limitations which will obscure the response. However there is a case for using the information when considering well managed irrigation schemes.

Water quality

The availability of water as a resource depends on quality as well as quantity. In general, temperature increase improves the self-purification processes in watercourses. However in the UK this is not the major consideration that determines water quality and the following issues should be addressed:

(i) Increased precipitation accompanied by an increase in the frequency and intensity of storms will increase storm run-off causing problems of reduced Phd.

(ii) Higher temperatures and the drying out of water-logged peat soils will lead to increased mineralization. Mineralization products (sulphur, ammonium, nitrate, phosphate and organic molecules) may be flushed out of the soils during re-wetting causing water quality problems.

(iii) Changes in wind direction may have a major effect on rainfall chemistry, altering the distribution and intensity of acid deposition and sea-salt events which have a major influence on the chemistry of many upland soils and freshwaters.

(iv) Reduced summer rainfall and increased evapotranspiration will change flow regimes, and sewage works may have greater difficulty meeting effluent dilution standards; storm drain effluents will be less effectively diluted.

(v) Agricultural responses to a warmer climate may include the increased use of pesticides, which may lead to an increased contamination of waters.

6. Regional changes in areas other than the UK

It is expected that the largest changes of temperature occurring from a doubling of carbon dioxide will be experienced in winter in the arctic and antarctic regions. Here temperature increases of up to 10°C or more might result. Although some small increases of temperature might occur in the tropics, the most striking change here would be a decrease in rainfall within the Inter-Tropical Convergence Zone of several mm/day. Similar changes might also occur in winter in the Mediterranean and southern USA regions.

Where precipitation is increased in winter the indications are that soil moisture would also increase. In summer there is likely to be a pronounced reduction in soil moisture attributed to earlier spring snowmelt and hence an earlier start to the drying out of the surface during spring. This earlier snowmelt would be more evident in high latitudes and the reduced precipitation in southerly latitudes.

The frequency of storms was mentioned in section 3. In general we might expect the storm belts to move southwards in the northern hemisphere and northwards in the southern hemisphere. Mid-latitude storms might on the whole weaken (mean wind speeds decrease), but extreme events occur slightly more often. Slightly increased sea surface temperatures in the tropics might increase the frequency of tropical storms by a few per cent.

Discussion

Raporteur: Professor C. G. COLLIER

Discussion opened with a contributor from the Climatic Research Institute, University of East Anglia informing the meeting of a project aimed at organizing the interface between the climate change impact community and the Hadley Centre for Climate Prediction and Research. The project would provide appropriate scenarios to those that needed them.

The question was raised as to whether it was correct that better regional models were needed to examine whether storminess would increase or not. It was felt that adequate models were already available to predict surges and waves provided a more reliable climate change scenario was to become available. However, another important factor seemed to be the track storms take. Any unusual tracks could result in even worse flooding. It was pointed out that there was no evidence indicating with any confidence, that storminess would increase around the United Kingdom. The evidence seemed to point to the fact that more of the precipitation was likely to be of a convective nature, suggesting a lower frequency of storm days, although when storms do occur they may be more intense.

On the question of work being done in the United Kingdom which might benefit developing countries, it was mentioned that Southern Africa was benefitting from Meteorological Office drought forecasts and that the Hadley Centre was carrying out research into seasonal forecasting which had particular applications in the Sahel and North East Brazil. Indeed, the level of skill obtained was better than expected on the basis of statistical analysis.

The French Meteorological Service is also developing a similar centre of expertise in Toulouse. Further improvements would only come with the refinement of the way in which the numerical models represented the interactions between the atmosphere, the oceans and the land surface. These refinements would be based upon the interdisciplinary work of many groups. Such co-ordination was already improving. The Institute of Hydrology is undertaking a project for ODA aimed at examining climate change scenarios to obtain some insight into the consequences for hydrology in developing countries. One problem already encountered was the difficulty of obtaining appropriate data sets, although recently a 1 km resolution global dataset had been obtained from the USA.

Finally, a contributor noted that the hydrological responses to changes in potential evaporation of 7% and 15% varied quite widely. Was it fair to conclude, he asked, that corresponding trends in the availability of water resources over the next forty years were 'lost in the noise' and could not be depended upon? There was some agreement with this point and the importance of obtaining more reliable information from the global climate models was noted.

The view of the National Rivers Authority

Dr C.J. SWINNERTON, Technical Director of the National Rivers Authority

1. Introduction

1.1 As guardians of the water environment, the National Rivers Authority (NRA) is responsible for water resources, pollution control, flood defence, fisheries, recreation, conservation and navigation in England and Wales.

1.2 Whilst the subject of this conference is primarily aimed at 'Engineering in the uncertainty of climate change', and some of the NRA's functions have a direct engineering context, this paper uses the opportunity of commenting on the impact of climate change across all of the NRA functions. Greater detail will be provided on the water resources and flood defence functions, as they have specific engineering interest.

1.3 As with many other environmentally 'hot' subjects, there are many conflicting views on the extent and timing of the impact of global warming. The NRA has based its consideration of the impact of climate change on the United Nations Intergovernmental Panel on Climate Change (IPCC) predictions. However, it has developed appropriate policies and response strategies.

1.4 It is important to recognise that there are direct impacts (e.g. sea-level rise) and indirect impacts (e.g. land-use change). The indirect impacts are even more difficult to predict than the direct.

1.5 It is not possible to prove conclusively that global warming is taking place, nor is it possible to be certain of its impact on the various elements of the hydrological cycle from both timing and geographical distribution viewpoints. However, the NRA takes the view that it is sensible to plan to be able to react to climate change in a pragmatic and cost-effective way as appropriate to each of its functions, where this is possible.

1.6 The NRA makes known to the climate modelling community, particularly the Hadley Centre for Climate Prediction and Research, its needs for information. Where appropriate, the NRA contributes to obtaining this data. The link between the modelling community and the impacts community is vital.

1.7 The following sections consider the potential impact of climate change in each of the NRA's functions, and relate this to the NRA's role in each of these and shows how the NRA is reacting to the potential in order to be able to continue to carry out its role effectively.

2. Water resources

2.1 The overall aim is to assess, manage, plan and conserve water resources, and to maintain and improve the quality of water for all who use it. Following from this aim, the strategic objectives most affected by climate change would be

(i) to plan for the sustainable development of water resources, taking account of the needs of the water environment and those of abstractors;

(ii) to investigate problems caused by authorised over-abstraction from water resources, and to implement a consistent approach to the alleviation of these problems.

2.2 Climate change due to global warming may affect the NRA's water resource activities in three main areas:

(i) changes in rainfall;

(ii) changes in temperature;

(iii) changes in evapotranspiration;

(iv) changes in demand.

2.3 These potential changes have to be considered alongside the changes which will occur regardless of climatic change. For example, demands for water are increasing anyway (see Fig. 1). Rainfall amounts vary significantly on an annual and monthly basis, as demonstrated by the 1989/90, 1990/91, 1991/2 event.

2.4 In March 1992, the NRA published a discussion document entitled 'Water Resources Development Strategy'. The basic aim of this document was to set the scene for the development of strategic water resources developments to meet future needs. Looking at the likely future demands, and comparing these with available yields from 'local' water resource developments, the conclusion reached was that there would be a need for the development of strategic options to transfer water from the wetter north

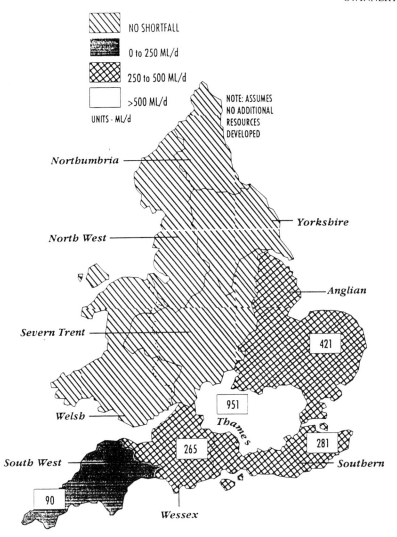

Fig. 1. Public water supply: shortfall in available reliable yield in 2021 based on average demand forecasts

and western parts of the country to the drier south and east. Figure 1 shows the estimated shortfall in water supply reliable yield in 2021, based on average demand forecasts, which in turn are shown in Fig. 2. A number of potential options have been identified with fairly crude cost comparisons with other ways of meeting demand.

2.5 The proposed way forward is to further evaluate the options and

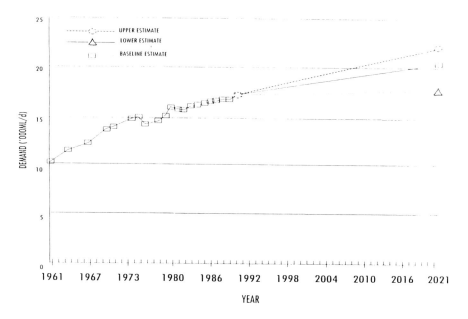

Fig. 2. Public water supply average demand projections to 2021

discuss with all interested parties to work towards an agreed overall framework for their development. Within this framework more local developments will need to be progressed.

2.6 Various interim steps will be needed, including reviews of demands by all potential categories of user, and the yield of existing sources and the yield of resource options. All these elements could be affected by climatic change.

2.7 Various studies have been carried out into the effects on water resources of changes in winter and summer rainfall and in evapotranspiration. The impact of such scenarios on surface water resources, groundwater resources and demands vary significantly. There have been relatively few studies into the potential impact on reservoir reliability. Similarly, little has been done to assess effects on groundwater recharge.

2.8 The overall conclusion is that there is no certainty of the effect on the various elements, and even if there was, the impact on a water resource system would need to be assessed by undertaking detailed modelling studies.

2.9 Other potential effects on water resources could be the increased saline intrusion due to sea-level rise, and the potential impact on surface water abstraction points near tidal limits.

2.10 The greatest impact on demands is likely to be for garden watering and demand for summer spray irrigation.

2.11 The majority of the former is supplied through the mains, and would reflect on the need for additional resource developments.

2.12 The majority of spray irrigation is in the south and east, and in many areas, it is already accepted that it is not possible to allow further direct summer abstraction from either surface or groundwater resources. The only way further spray irrigation needs can be met is for winter storage to be provided so that abstraction during the summer months is not direct from the river. Recent studies have indicated that a typical storage requirement at 3 m depth could require 4% of a land holding to be sacrificed to provide adequate winter storage. An impact of climate change could be to require a somewhat increased percentage of land requirements for the provision of storage.

2.13 Other impacts on demands could be a variation in the frequency of burst pipes and impact on the demand for cooling water.

2.14 From the above it is clear that potential climate change could affect many of the elements involved in the planning of future water resources. However it is difficult to predict the scale and timing of such effects.

2.15 In order to be able to determine potential effects more reliably, the NRA requires models which are capable of simulating the effect on:

(i) River flows and groundwater under future climatic conditions;

(ii) Further advice on whether climatic change is occurring (the NRA will continue to rely on IPCC information);

(iii) A set of agreed climatic change scenarios and agreed methodologies for expressing resource availability in the light of climate change.

All these elements are receiving attention.

2.16 Whilst this work continues, the adopted approach is that forward planning must be done based on the best information available at present, accepting that whatever the actual effects of climatic change, the direct impact would be on the timing and total scale of development needs. Such an approach is needed due to the uncertainty in the process of demand forecasting as illustrated by Fig. 4, which compares the NRA's current baseline forecast and that of the Water Resources Board. Providing sufficient potential resources are identified, a flexible approach will need to be taken on the timing of these developments.

Fig. 3. Major transfer options: examples of inter-regional transfers

2.17 If the impact of climate change was so severe that the conventional types of resource developments are not feasible, it is suggested that there would be tremendous impact on life in the UK overall for reasons other than water resources. If this was the case, alternative types of resource development such as desalination would need to be considered. Whilst this option is expensive in comparison with the more conventional resources at present, the overall impact of severe climatic change on the UK (and world) economy could change this situation such that the cost of moving to desalination would be a relatively small component of the total impact.

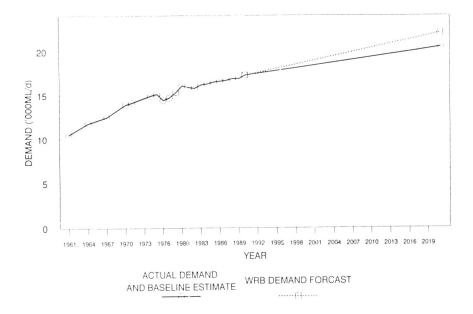

Fig. 4. Public water supply demand projections to 2021

2.18 Mention was made earlier of the NRA's work to alleviate low-flow problems due to over-abstraction. Early on, the NRA identified 40 catchments which were felt to be affected by over-abstraction. Of these, 20 priority rivers were identified and investigations have been carried out on these. (See Fig. 5).

2.19 On some of these, such as the river Ver, the investigations have led to solutions being implemented. On others, investigations are continuing.

2.20 The impact of climate change could be to increase the number of catchments in which abstractions are too large a proportion of available resources, due to reduction in rainfall and recharge to resources. The potential cost of solving existing problems is very large and any significant increase in number of catchments would probably be beyond the current funding arrangements for water resources to cope. This is especially true as there are far more rivers than the original 40 identified which are already affected by over-abstraction and need investigation and solution.

3. Flood defence

3.1 The key aims of the NRA in flood defence are:

(i) to provide effective defence for people and property against flooding from rivers and the sea;

(ii) to provide adequate arrangements for flood forecasting warning and response to flood events.

3.2 Objectives aimed at achieving these aims which are potentially most likely to be affected by climatic change are:
(i) determination of maintenance works and capital works to provide effective defence;
(ii) influence development control in flood plains;

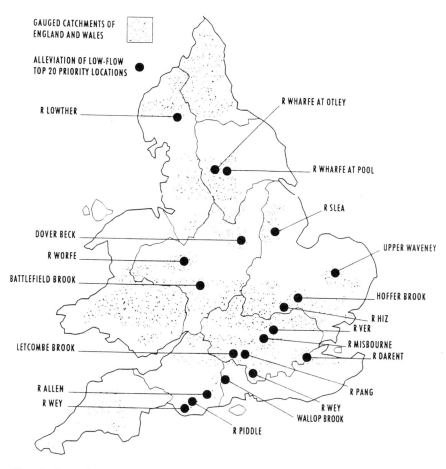

Fig. 5. Low flow priority locations and extent of gauged catchments

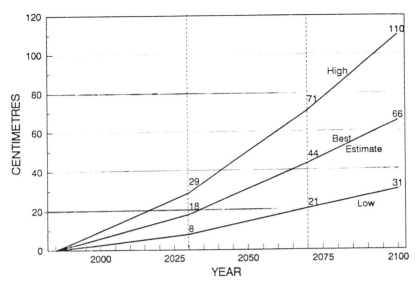

Fig. 6. Predicted future global sea level rise according to the IPCC "business as usual" scenario

(iii) carry out and support relevant research and development.

3.3 Of all the NRA functions, it is probably flood defence in which most attention has been given to climatic change, particularly in relation to sea-level rise.

3.4 As stated earlier, the NRA accepts the IPCC predictions for climate change and for sea-level rise, a rise of 180 mm between 1990 and 2030 is taken for planning purposes (see Fig. 6). A 480 mm increase is taken for the period 2030 to 2100. The net change in sea-level needs to combine the effects of sea-level rise, with any changes due to tectonic or isostatic effects. The change due to this varies from −0.5 mm per year for the Northumbria and North West regions of the NRA to +1.5 mm in the Southern and Thames regions. The net effect ranges from an increase in sea-level from 4.0 mm per year to 6.0 mm/year between 1990 and 2030, and between 6.0 mm/year and 8.5 mm/year for the period between 2030 and 2100.

3.5 These figures (shown in Table 1) are used unless results from site specific investigations are available.

3.6 The approach followed by the NRA is to include these net changes in sea-level for use in the planning of future sea and estuarial defences. This does not mean that defences will be built to a height which would allow for this over the planning period, but that designs are such that

25

Table 1. Table of allowances

REGION	TECTONIC OR ISOSTATIC CHANGE (Selected From Fig 2) MM/YEAR - UPWARD MOVEMENT OF LAND + DOWNWARD MOVEMENT OF LAND	IPCC SEA LEVEL CHANGE "BUSINESS AS USUAL, BEST ESTIMATE"	
		1990 - 2030 180 MM 4.5 MM/YEAR	2030 - 2100 480 MM 7.0 MM/YEAR
		Combined Effect of Climate Change & Tectonic Change	
Anglian	1.5	6.0	8.5
Northumbrian	- 0.5	4.0	6.5
North West	- 0.5	4.0	6.5
S-Trent	0.5	5.0	7.5
Southern	1.5	6.0	8.0 *
South West	0.5	5.0	7.5
Thames +	1.5	6.0	8.0 *
Wessex	0.5	5.0	7.5
Welsh	0.5	5.0	7.0 *
Yorkshire	0.5	5.0	7.5

+ Estuary allowance. Not considered applicable in other estuaries u/s Schedule 4, Coast Protection Act 1949, Boundary.
* For period 2030-2100 amended rate reduced by 0.5mm. Current uncertainty about range of tectonic effect justifies reduction of long-term allowance.

PHASES 1, 2 AND 3 COMBINED

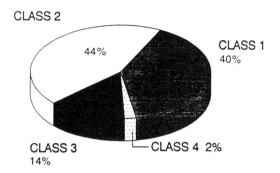

PHASES 1, 2 AND 3

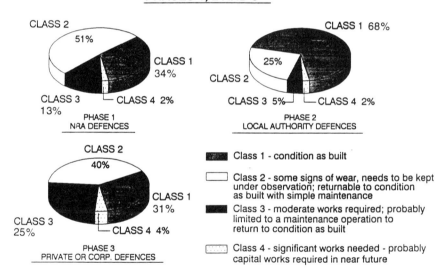

Fig. 7. General condition of defence elements

foundations are designed and constructed so that defences can be raised to the required height as and when it is shown to be necessary.

3.7 MAFF who provide support for sea-defence capital works endorse this approach and will grant aid to such schemes subject to them being cost-effective.

3.8 During 1990 and 1991, the NRA carried out a survey of all sea-defences in England and Wales, regardless of ownership. The results of the survey were published in May 1992 in a document entitled 'The NRA Survey of Sea-defences'. The results on the general condition of the defences are summarised in Fig. 7.

3.9 The survey provided for the first time a comprehensive survey of all the sea-defences in England and Wales and produced information on the condition of the defences, the residual life of the defences and the priority to be given to the defences depending on the type of property and land protected. This information provides a sound basis on which to plan future works. Where NRA defences were in need of moderate or major work to bring them back to an 'as-built' condition the necessary schemes have been included in the capital programme for 1992/3.

3.10 The survey has been extended to cover estuarial defences and the results of this phase of the survey will be published in late 1992.

3.11 The availability of this data and the agreed policy on allowance for sea-level rise provides a sound basis to ensure that the NRA's defences are kept in good order. Expenditure on this area of work is about £50M per year, and this amount is likely to gradually increase in future years.

3.12 The NRA has an overall supervisory duty over all sea-defences, and where necessary the NRA is in discussions with the owners of sea and estuarial defences in order that the necessary work is carried out.

3.13 A rise in sea-level would give a greater depth of water at defences, which would result in increased wave height. Due to the non-linear relationship between wave height and energy (e.g. wave height factor of 2 would mean about 6 times the energy) more substantial as well as higher defences would be needed.

3.14 The frequency of large waves may increase due to sea-level rise, but at present it is difficult to predict this accurately.

3.15 Climate change could have an impact on fluvial defences but far less attention has been paid to this. Impacts could be:

(i) changes in prolonged heavy rainfall amounts and changes in antecedent moisture conditions;

(ii) changes in intensity and frequency of local rainstorms;

(iii) changes in run-off from snow melt.

3.16 As for the changes in the rainfall pattern's effects on water resources, these changes cannot be predicted with any certainty, but the NRA will need to stay aware of developments so that it can react as necessary.

3.17 Returning to sea and esturial defences, it is considered that if the IPCC best estimates occurred, the current policy of providing defences where the benefit/cost ratio exceeds unity, it will still enable property and people to be protected as now. If much more extreme rises occur, this may not be possible. Alternative scenarios would have to be given greater consideration than now. These would include:

(i) abandon sea-defences and retreat inland;

(ii) maintain current line of defences but with a lower standard of protection.

3.18 The 'do nothing' and retreat options are already considered for individual defence schemes, and the implemented scheme has to be cost effective. Whilst it is considered correct that these options are included, any comprehensive policy to retreat and sacrifice massive areas of land would have implications that such a decision could not be taken by the NRA alone. It would certainly involve Government.

3.19 The NRA is consulted on development proposals in flood risk areas and advises against developments which will increase the risk of flooding, unless appropriate compensatory arrangements can be provided. However, the advice is not binding and the final decision is taken by the Planning Authority. Unless climatic change is much greater than current IPCC predictions it is thought unlikely that the impact on the NRA's development control activities will be great.

3.20 Whilst the NRA has developed the above approach on climatic change in relation to flood defence, the following is still required:

(i) greater reliability in predictions of climatic change on sea-level rise, surge effects and impact on rainfall patterns etc.;

(ii) information on the relative sensitivity of different catchment types to changes in rainfall characteristics;

(iii) procedure to estimate the effects on flood characteristics of different rainfall scenarios;

(iv) methods for estimating flood risk against climate change;

(v) information on flexible design techniques of flood defences;

(vi) information on possible impact of changed river flow on rivers.

(vii) impacts of sea-level rise on sediment transport particularly at the coast, but also in rivers if flow patterns alter.

4. Pollution control

4.1 The key aim of the NRA on pollution control is to achieve a continuing improvement in the quality of rivers, estuaries and coastal waters through the control of pollution. Objectives which are targeted towards this aim, which would be the most affected by climate change are:

(i) advising the Secretary of State on the setting of Statutory Water Quality Objectives (SWQOs) and standards;

(ii) control of point discharges or diffuse pollution so that the SWQOs will be achieved.

4.2 In 1991 the NRA began preparing proposals for SWQOs and standards. These will include a set of water use categories, each with appropriate water quality standards, and a revised water quality classification scheme. The system will also include biological criteria override so that due importance can be given to the biological requirements of waters.

4.3 Climate change would affect water quality and pollution both directly and indirectly. Direct changes would primarily be due to changes to the hydrological processes as for water resources. Indirectly, the impacts would be due to changes in land use and the subsequent impact on water quality.

4.4 It is convenient to consider the effects on water quality of upland and lowland waters separately.

4.5 Lowland rivers tend to be slower flowing, with low turbulence and dissolved oxygen generally below saturation. They also tend to have high nutrient concentrations. They often provide water for abstraction, but also are used for the removal of effluents.

4.6 River water quality problems in lowland rivers can arise from:

(i) high concentrations of a wide range of pollutions, including nutrients and toxic chemicals or

(ii) low concentrations of beneficial elements, notably dissolved oxygen.

4.7 Climatic change could increase concentrations because of reduced flows, or increase survival time of pathogenic bacteria because of increased temperatures. Changes in land use which might include increased pesticide or fertilizer use would also impact on water quality.

4.8 A scenario of lower flow, higher temperature and higher nutrient concentrations might promote algal growth, and result in de-oxygenation. But there is uncertainty whether recent toxic blue-green algae blooms result from natural processes or as a result of increased agricultural pollution.

4.9 Overall, the impact on water quality must be carefully considered for each pollutant and generalities are difficult.

4.10 For uplands waters, the quality is essentially controlled by natural processes within their catchments. The likely main direct impact of climate change would be an increase in nitrate concentration through mineralisation of organic nitrogen in the soil, brought about by increased temperature. The result would be an increase in surface water acidity. Lake productivity might increase but eutrophication is not likely to be a problem.

4.11 Under possible scenarios of lower summer rainfall, higher temperatures but increased winter rainfall would result in greater water discoloration, causing enhanced flushes of colour.

4.12 In relation to estuaries and tidal waters, there are 3 issues which affect water quality. These are:

(i) presence of saline water;

(ii) distribution of sediments and their movement;

(iii) pollutants flowing into the tidal reach from the non-tidal river.

4.13 Sea-level rise could increase saline intrusion. Changes in sediment movement could cause navigation and quality problems. Reduced flows would reduce flushing out of pollutants and sediments. Increased temperatures could speed up biochemical processes.

4.14 Increased temperatures might improve the quality of coastal bathing waters, but could increase the demand for bathing waters.

4.15 In urban areas possible increased frequency of intense rainstorms could cause more frequent storm overflows to the detriment of water quality.

5. Fisheries

5.1 The key aim for the NRA on fisheries is to maintain, improve and develop fisheries

5.2 Aspects of fisheries work which could be affected by climatic change are:

(i) changes in flow and temperature pattern which would affect distribution of the various species for fish; e.g. the southerly limit of salmonids could move further north.

(ii) changes in migratory habits of salmonids.

5.3 The fisheries resource is managed for the following reasons:

(i) it is a wildlife resource in need of conservation;

(ii) it is a commercially exploited resource;

(iii) it is a resource exploited for recreation.

5.4 Salmon and trout fisheries predominate in the North and West, and trout are important in many chalk streams in the South, East and Central England.

5.5 Climate change could result in higher temperatures, changed river flow regimes, changes in water quality and changes in ocean circulation. In turn, these changes could impact upon fish physiology, fish habitat, food resources and migration patterns.

5.6 Increases in winter temperature could affect the spawning and embryonic development of a number of fish species in the UK. Eggs and embryos of most fish tolerate a much narrower range of temperatures than those tolerated by juveniles and adults. Spawning performance of most trout and salmon populations should not be adversely affected, but increased spring temperature could induce early emergence and perhaps a shortage of food would exist at this time. Roach, bream, carp and perch spawn earlier in the year in Southern Europe, and they would almost certainly adapt.

5.7 As an example of temperature effects on lakes, the charr in Windemere could be confined to a very narrow range of depths, with resultant feeding difficulties.

5.8 In addition to the impact in salmonids, another vulnerable fish is the native brown trout. Growth rate would be much slower than today. At temperatures above 14°C, the trout require large amounts of food to compensate for the energy lost in excretory products.

5.9 The basic controls of fish habitat are water velocity and depth, channel substrate in channel vegetation, bank characteristics and vegetation and water quality. Different fish have different habitat requirements.

5.10 Warmer summers and more frequent droughts would affect river fish, especially young fish. This would be exacerbated by lows due to lack of rainfall.

5.11 Droughts affect the survival of young fish in upper reaches, and may also discourage migratory fish from moving upstream to spawn. However, the number of salmonids moving upstream in the autumn could increase over present numbers if summer droughts were followed by heavy rains, although many other factors influence the runs of migratory fish.

5.12 The routes for migratory fish, and the food and feeding habits of fish could be affected by changes in ocean circulation.

5.13 Changes in river flows and qualities will affect fish food sources, particularly the abundance of invertebrates, but the impact is very difficult to access.

5.14 The NRA will not be able to prevent the fish population changing in reaction to climate change. It might however, need to consider stocking policies appropriate to the changed circumstances.

5.15 There may be a need to react to more frequent fish kills because of changed water quality and increased water temperatures.

5.16 Disease may be more prevalent and there could be different diseases. Hence, there may be implications for the management of disease outbreaks.

5.17 Exotic species of fish could move into UK coastal waters from the currently warmer areas.

5.18 To help the NRA react to the potential impact on fisheries, the NRA needs information on: potential changes in water temperature; the impact of changes in river flow regimes and water quality on habitat suitability and fish population; the effect of changes in water temperature on fish physiology and disease; the changed sensitivity of fish to pollution in higher temperature waters

6. Recreation and navigation

6.1 The key aims for recreation and navigation are:

(i) to develop the amenity and recreational potential of waters and land under NRA control;

(ii) to improve and maintain inland waterways and their facilities for use by the public where the NRA is the Navigation Authority.

6.2 Water based recreation takes place both on and in water. Changes in water quality due to climatic change would affect recreation as could changes in temperature and changes in the rainfall patterns.

6.3 Changes in the frequency and scale of toxic algal blooms would be of particular significance.

6.4 Changes in rainfall pattern to the detriment of water flows could make it more difficult to maintain water levels adequate for navigational and other recreational purposes. Navigation on some waterways is already restricted in dry summers because of lack of water, and this would occur on more systems. Sedimentation changes could also make it more difficult to maintain adequate water depth, although increased winter rainfall could increase flood flows and resultant scouring.

7. Conservation

7.1 The aim of the NRA for conservation is to conserve and enhance the wildlife, landscape and archaeological features associated with inland and coastal waters.

7.2 Climate change resulting in increased temperature and increased in soil moisture deficit would lead to changes in species composition of ecosystems. Even changes in average temperatures as small as 1°C could significantly alter species compositions in statutory protected sites in the UK. Species at greatest risk are those which are already at the limit of their distribution, or are in isolated communities.

7.3 An indirect impact could occur due to changes in agricultural and land use practices, and the activities of other NRA functions.

7.4 Conservation interests in coastal ecosystems could be affected by sea-level rise and changes in tidal regimes. The distribution of mud flat invertebrates would be affected by changes in tidal range and rate of erosion.

7.5 In general, a rise in sea-level would result in the cities becoming poorer and less diverse. This in turn would greatly reduce the feeding potential for many bird species.

7.6 It is considered impractical to maintain a particular ecosystem against climatic change. The NRA would therefore need to deal with particularly exposed communities.

7.7 To help develop appropriate strategies, the NRA needs information on:

(i) ecosystems elements most sensitive to climatic change and sites at greatest risk;

(ii) the sensitivity of important species and communities;

(iii) rate of climate change;

(iv) likely impact on changes in land use;

(v) a prioritization of what needs to be conserved.

8. Key impact areas

8.1 The NRA directly and indirectly supports a range of research and development targeted at the areas of work carried out by the NRA to provide information on the key impact areas. Many of the areas of R&D are related to providing the required information to better understand and to react to climatic age.

8.2 Key impact areas are:

(i) impact on evapotranspiration;

(ii) changes in groundwater recharge;

(iii) implications for the operation of integrated water management systems;

(iv) the importance of the 1988 – 1992 drought (or whenever it ends);

(v) changes in land use due to climate change;

(vi) critical thresholds for specific water management systems;

(vii) changes on fluvial flood characteristics;

(viii) combining historical information with scenarios of future climate;

(vii) changes on fluvial flood characteristics;

(viii) combining historical information with scenarios of future climate;

(ix) changes in agricultural practice with implications for agricultural pollution;

(x) impact on water temperature;

(xi) impact on fish habitats of changes in flow regimes and water quality;

(xii) impact of changes in water temperature on fish physiology and disease;

(xiii) changed sensitivity of fish to pollution in higher temperature water;

(xiv) sensitivity of important river ecosystems;

(xv) expressing hydrological characteristics against a changing climate;

(xvi) effects on recreational potential of water courses.

(xvii) constraints on coarse fish populations.

(xviii) sea trout studies.

8.3 The NRA is co-operating with others to identify and agree appropriate R&D programmes to provide the necessary information on these key impact areas.

9. Conclusions

9.1 All of the NRA's functions would be affected by climate change to varying degrees. The IPCC predictions are used by the NRA.

9.2 For most functions the current predictions of climate change are not sufficiently reliable to know with any certainty what the effects might be.

9.3 In sea-defences it has been possible to develop a policy which helps ensure that defences are designed and built so they can be increased in height, if necessary, in the event of proven sea-level rise.

9.4 The variables that would be affected by climate change which could impact upon most of the NRA's activities are rainfall, flow regimes, water temperatures and weather patterns generally. Until better predictions are available, it is not possible to be precise in planning for the future.

9.5 It is essential that the NRA takes a responsible attitude to climate change, but it must also be realistic and not over-react. Part of this responsible approach is to ensure that appropriate R&D is carried out to fill the existing gaps in knowledge which exist across all NRA functions.

Acknowledgements

This paper has used information from a variety of sources; in particular from relevant NRA Research and Development Reports, the NRA Corporate Plan and various NRA policy documents.

The author acknowledges the work done by others in producing the other documents, and expresses his appreciation. Whilst the opportunity to comment has been provided to some of these other NRA staff, the author apologises for the fact that not everyone could be consulted.

Water resource planning in the uncertainty of climatic change: a water company perspective

B. A. O. HEWETT, Group Director of Corporate Affairs, Southern Water Plc, C. D. HARRIES, Deputy Head, Regulation and Planning, Southern Water Services Ltd, and C. R. FENN, Manager, Hydrology and Modelling Southern Science Ltd

Introduction

In July 1990, during the second year of a groundwater drought which has yet to abate and prior to the publication of the Report of Working Group 1 of the Intergovernmental Panel on Climate Change (IPCC) (Houghton *et al.*, 1990) and the First Report of the United Kingdom Climate Change Impacts Review Group (UKCCIRG) (Department of the Environment, 1991), Southern Water Services Ltd instructed Southern Science Ltd to undertake three climate change studies on its behalf:

(a) climate change due to the enhanced greenhouse effect: implications for the interests of Southern Water Services

(b) sea level rise, and its implications for the interests of Southern Water Services

(c) patterns of recharge and recession of groundwater levels in the Chalk aquifers of south-east England

The commissioning of these studies reflected an early appreciation that climate change, were it to occur, would impact directly upon the water supply and wastewater disposal functions of Southern Water Services. It also reflected a wariness born of the effect upon resources and demands of a (then four season) sequence of dry winters and summers. By July 1990, demand restriction measures were in place, for the second successive year, in the Kent and Sussex Divisions of the Company, and water managers were posing questions such as: Is this due to climate change? Is this what we can expect under climate change? Will these events occur more frequently under climate change? If climate change happens, will we be able to cope?

Then, as now, there existed a strong scientific consensus that the emission of radiatively-active gases is generating an enhanced greenhouse effect (EGE) which has the potential to cause global climate change of unprecedented rate, scale and magnitude; but considerable uncertainty as to the likely changes in climate which any given region is likely to experience over time.

The purpose of the three climate change 'scoping studies' referred to above was two-fold: (a) to determine best estimates of likely changes in climate in south-east England, and best assessments of the implications thereof; (b) to determine how the Company should respond. Based upon the results of these studies, and mindful of the increasing scarcity and tighter regulation of water resources in Southern England, Southern Water Services decided that climate change considerations should be built into its strategic planning policy. This paper outlines how this has been done, and how the problems and uncertainties associated with climatic change are being addressed.

Southern Water Services' response to the uncertainties associated with climate change

In the UK context, the Southern Water region is in the front-line on the climate change/limited water resources issue. It is one of the regions most badly affected by the sustained drought of 1989–?. If climate change occurs, the Southern region will be one of those in which the forecast changes in temperature, precipitation, evaporation and storminess will be at their greatest. Sea level rise, and the attendant threats of saline intrusion into aquifers, estuaries and rivers will also be at a UK maximum along a coastline where land subsidence (tectonic plus isostatic) and eustatic rise may, in some areas, contribute to net sea level rise in almost equal measure.

The catalogue of potential effects of scenario climatic changes on water resources and the demands upon them identified in the scoping studies conducted by Southern Science during late 1990 and early 1991 (Fenn, 1990; Fenn, 1991; Munnery, 1991) is by now a familiar one. That the forecast changes in air temperature, rainfall and evaporation will cause changes in raw water availability (via their effects on river flow, reservoir refill and groundwater recharge regimes) and demands upon resources (via their effects on patterns of domestic, agricultural and industrial consumption) is not in doubt; but the magnitude and even the direction of the various changes are by no means clear) e.g. Arnell *et al.*, 1990; Cole *et*

39

al., 1991). Part of the uncertainty derives from that surrounding the scenarios which are used for 'perturbed state modelling'. Global Circulation Models (GCMs) provide inconsistent and spatially coarse estimates of changes in air temperature and rainfall, so it follows that regional scale estimates of hydrologically-significant variables (e.g. rainfall, evaporation, soil moisture content) are uncertain; there is thus no definitive scenario, but rather a plethora of scenarios of uncertain validity. Further uncertainty results from the highly sensitive gearing of hydrological response to climatic forcing. Arneli *et al.* (1990) and Arnell (in press) describe how climatic changes are amplified as they are filtered through the hydrological cycle; how response varies according to catchment properties; and how the response of river flow to a change in rainfall is sensitively dependent upon the assumed change in evapotranspiration.

The National Rivers Authority's (NRA's) response to this uncertainty is declared in their Water Resources Development Strategy Discussion Document of March 1992:

"The possibility of climatic change adds an element of uncertainty to the estimation of future demand. To date there is insufficient knowledge to enable quantitative incorporation of the impact of climate change on demands for public water supply. However, it is expected that the main impact would be in relation to garden watering." (p.3)

"Various climate change scenarios have been postulated but as yet there is insufficient evidence to have confidence in any particular scenario. However, it is generally expected that if climate change is going to impact upon the availability of water resources it will take place gradually and enable sufficient time to respond to the change." (p.5)

Southern Water Services' response to the problems and uncertainties of potential climate change, in contrast, has been to accept the likelihood of climate change as a working hypothesis, and to build climate change considerations into its planning policy. This decision reflects the high impact which forecast changes in climate could have on the Company's operations. It also reflects a belief that a 'wait until we have clearer evidence' attitude is not an appropriate one for a company with duties to provide service to prescribed standards, whatever the operating circumstances. Any change arising from the EGE will be superimposed upon natural variability in climate. Because the EGE is likely to take effect gradually does not necessarily mean that there will be time to respond to the problem as and when it occurs; 'natural variability' in climate is considerable, and displays some propensity for clustering, so it will never be easy to 'spot' climate change. Since the signal to other effects ratio of the EGE is likely to remain indistinct for many years to come, the current and future climatic — and thus hydrological — situation in which a water company must conduct its business is one of uncertainty.

In striking a balance between evidence and consequence in respect of climate change, Southern Water Services have elected to adopt what has been termed a 'no regrets' policy (Owens, 1991); i.e. neither embarking upon a policy which it would otherwise not have done, nor embarking upon a policy without taking climate change consequences into account. In practical terms, the company's response to climate change has been: first, to commission scoping studies to identify the range and impact of possible effects; second, to develop a resource planning model to enable observed (whether climate change induced or otherwise) variations in resources and demands to be incorporated, analysed and acted upon; third, to conduct scenario modelling and sensitivity studies designed to identify best practice in the prevailing climate of uncertainty. These responses have been implemented according to a phased strategy designed to enable the company to discharge its functions in the presence of regulatory require-ments, customer/shareholder requirements and climatic unsteadiness.

Incorporation of climate change issues into Southern Water Services' strategic planning policy

Post-privatisation, Southern Water Services' resource planning policy has had to adapt and become more flexible and sophisticated, in order to take account not only of the effects of potential climate change, but also:

- the new post-privatisation regulatory regime
- the increasing scarcity of water resources in SE England
- the severe drought of 1989–
- the need to provide effective services to the customer at least cost (i.e. the balancing of the customer/shareholder relationship)

Regulatory influences on resource planning in the post-privatisation era

The very nature of the water business has meant that water resource planning has always taken a central role in the overall planning process. In the last 20 years, legislative and other changes have sharpened the planning focus at a company level whilst introducing an increasing spectrum of broader external issues which need to be taken into account. Water resources, once the preserve of the small water companies, are now a national issue with many stakeholders all seeking to steward this important resource.

Privatisation of the water industry in 1989 gave rise to a new regulatory regime which now has an important influence on the water resource planning process. From a water resources viewpoint there are two key bodies:

1. The Office of Water Services (OFWAT). OFWAT represents the customer's interests. It regulates the amount that companies can charge customers and the service provided by this income. The ability to finance new resources is therefore influenced by OFWAT.
2. The National Rivers Authority (NRA). The NRA is the environmental steward, charged with licensing and managing the resource environment. The NRA therefore controls a company's ability to develop a new resource.

After some three years of the new economic regulatory regime, OFWAT have directed that water companies should prepare a fresh appraisal of their investment projections as part of a broader resetting of the financing regime. The whole process is called the Periodic Review. The reassessment of investment needs is called the Asset Management Plan (AMP2).

OFWAT's aim is to have a new price cap regime in place with effect from 1st April 1995, so that charges to customers can reflect the latest obligations being imposed on the industry many of which emanate from the European Community.

The investment projections in AMP2 look 20 years ahead and, as such, could well be influenced by climatic change. It will be important for Southern Water Services, located in the drought-prone South East of the country, to be clear about future resource risk and to make sure that appropriate financial provision has been made for those risks to be covered by the company over this time period.

As part of their regulatory regime, OFWAT have also been reconsidering the way in which they monitor companies' performance. The two current measures for water resources concentrate on:

1. The long-term reliability of a company's resources relative to the demands on these resources. This indicator is known as DG1 — Water Resource Reliability Factor

2. A more immediate measure of the extent to which a company imposes a hosepipe ban, has water restriction publicity campaigns or introduces stand pipes. This indicator is known as DG4 — Water Use Restrictions.

Individually, DG1 and DG4 provide a measure of how well a company is performing under the terms of the regulatory contract. It is this contract which allows a company to raise monies from customers in return for achieving specified output services. DG1 and DG4 however, do not provide a clear set of priorities. DG1 deals in hydrological return events; in Southern Water Services' case a one in 50 year return period event has been used to assess source yields. DG4 however, looks at the frequency of restrictions imposed as a result of immediate resource shortages: hosepipe bans no more frequently than once in 10 years, publicity campaigns no more frequently than once in 20 years and standpipes no more than once in 100 years.

In parallel with this, the NRA have been developing their approach to resource licensing. The NRA's evolving policy objectives have been defined in a series of documents issued at regional and national level. The well-publicised issue of low flows arising from historical over-licensing of public water supply (PWS) sources is central to their deliberations, and has given rise to a preferred policy of limiting further resource developments to catchments best able to sustain abstraction. inter-basin transfers and demand management strategies are seen as the principal instruments for achieving this objective (e.g. NRA Southern Region Draft Water Resources Strategy November 1991 (NRA, 1991); Water Resources Development Strategy Discussion Document, March 1992 (NRA, 1992)).

Evolution of Southern Water Services' resource planning strategy

Against a background of increasing regulation, and increasing uncertainty as to the reliability of resources to meet unrestrained demands under a scenario of climatic instability, resource planning in Southern Water Services has followed a series of phases which will culminate in the production of costed resource projections for AMP2.

Phase I: (1989). In 1989, Southern Science were commissioned to review the resource situation in each of Southern's water resource areas and to draw up a series of options for resource development which could be considered. This first comprehensive documentation — The Water Resources (Giles 1989) — provided a foundation for subsequent deliberations in the light of the drought which followed so closely on the heels of the privatisation process itself.

Phase II: 1990. The drought of 1989–? prompted Southern Water Services to commission the climate change scoping studies described earlier. In late 1990, following the first year of reporting to OFWAT, it was apparent that the relationship between the two level of service indicators DG1 and DG4 was not clear. Future resource planning needed to identify the most onerous of the two and, at the same time, explore the effect of different return period scenarios. Southern Science were again commissioned to study the links between the two and to make recommendations as to how this issue might be progressed. There was also a need to identify the appropriate factors of safety within the DG1 calculation itself.

Phase III: (1991). Southern Science's report on the calculation and relationship of the level of service indicators DG1 and DG4 (Giles, 1991) opened the way to updating and extending the original Water Resources Review. Southern Science were commissioned to develop a methodology for assessing resource area demands, resources and levels of service in design drought events of 1:10, 1:20, 1:50 and 1:100 years return periods; to develop a working model for storing, retrieving and manipulating the relevant data; and to use the methodology and model to update and extend the demand and resources estimates given in the first version of the Water Resources Review. The resulting Water Resources Review version 2 (Fenn, 1992) will form the basis of the company's investment projections for the AMP2.

Phase IV: 1992. In the fourth phase of the evolution of Southern Water Services' resource strategy, Southern Science have been asked to model possible effects of climatic change on critical resources. This modelling work will be taken into account in the costed, most probable scenario adopted for submission as part of the AMP2.

Climate change considerations have been explicitly taken into account in the second, third and fourth phases of this strategy. A cornerstone of the company's policy in respect of climate change has been to develop working tools to enable the overall impact of given changes in resources or demands to be determined before embarking on scenario studies designed to estimate the details of the changes which may occur in particular resource or demand components. Thus, sensitivity studies have preceded scenario studies for two main reasons. First, the ability to reckon the consequences of any specified set of changes enables the company to define its vulnerability to climate change induced effects, and allows it to respond accordingly. Second, in circumstances of uncertainty as to scenario validity and data stationarity, it makes good sense to lay emphasis on developing frameworks for computing the effects of changes rather than on deriving detailed estimates of the changes which could occur (since, in all probability, these will need to be revised again and again).

Development of Southern Water Services' strategic planning model and its use as a general tool for climate change sensitivity studies

The development of a spreadsheet model for computing demand, resource and level of service forecasts for a range of design drought frequencies (1:10, 1:20, 1:50, 1:100 years return periods) may not at first sight appear to be particularly significant with respect to establishing the effects of climate change in a resource planning context. The model has

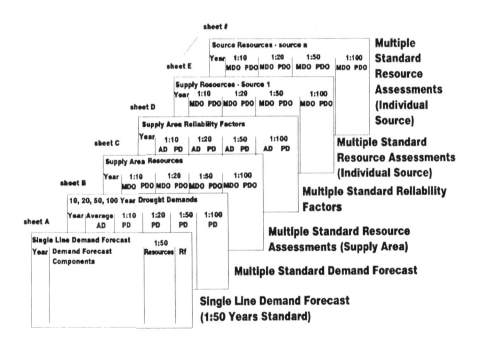

Fig. 1. Schematic of the stacked spreadsheet system developed for computing demand, resource and level of service forecasts, and for conducting what if? simulations.

been constructed, however, so as to enable it to act not only as an information system, with storage–retrieval and dynamic updating facilities, but also as a working tool for resource planning strategy development: it defines 'now' and 'future' situations, and enables 'what if' modelling to be undertaken quickly, easily and accurately. Climatic change is one of the what ifs which may be evaluated.

Figure 1 shows the structure of the spreadsheet model developed. The model was built with version 3 of the Lotus 1-2-3 package (Lotus Corporation, 1990). A stacked spreadsheet (i.e. a spreadsheet file consisting of multiple sheets incorporating information and calculation flows across, down and between separate sheets) has been created for each of the resource areas in the four divisions of Southern Water Services (Kent, Sussex, Hampshire and Isle of Wight). The spreadsheet is structured as follows:

- Sheet A: Component-based single-line demand forecast;
- Sheet B: Average demand (AD) and peak demand (PD) forecasts for design droughts of 1:10, 1:20, 1:50 and 1: 100 years return periods;
- Sheet C: Resource area minimum drought outputs (MDO) and peak drought outputs (PDO) during design drought of 1:10, 1:20, 1:50 and 1:100 years return periods;
- Sheet D: Level of service factors (for MDO/average demand and PDO/peak demand conditions) during design droughts of 1:10, 1:20, 1:50 and 1:100 years return periods;
- Sheets E, *et seq.*: MDO and PDO for an individual source during design droughts of 1:10, 1:20, 1:50 and 1:100 years return periods.

Each spreadsheet has been written in such a way that changes in any of the data input fields (e.g. any component of the demand forecast, any MDO or PDO) automatically flow through the calculation system to produce updated calculations of resource area-scale demands, yields and levels of service. Given that all data on demands and resources are estimates, uncertainties in end estimates of demands, resources and reliability factors are defined by confidence bands around best estimates. The uncertainty bands may be fixed according to any required combination of modified assumptions (e.g. 1, adopting, say, 1% per annum growth as the best estimate of per capita domestic consumption, and 0.5% p.a. and 1.5% p.a. as lower and upper limits; e.g. 2, assuming that resource estimates are in error by some fixed amount or percentage, or that the 1:50 year estimate should in fact be a 1:20 years or 1:100 years estimate).

The model can be used to define which of the various design standards (e.g. the 1:10 years hosepipe standard, the 1:50 years resources standard) is (are) critical for strategic planning and investment purposes. Thus it can be used to determine policy decisions on the basis of current best estimates of demands and resources. Similarly, by varying demand and/or resource conditions to correspond to any of a number of climatic change scenarios, the model can be used to

- simulate the effect of any plausible set of climate change induced changes in source drought outputs, per capita domestic consumption, industrial consumption etc.,
- determine the sensitivity of a resource area to any assumed set of resource and/or demand changes,
- explore the effects of different investment decisions,
- assess the company's performance under a range of assumed scenarios.

The model thus enables the company to determine its vulnerability to climate change effects of specified type, direction and degree, and to judge the consequences of any given change in advance of that change occurring. Thus it facilitates proactive as opposed to reactive planning.

Site-specific scenario studies

The fourth phase of Southern Water Services' resource planning strategy includes examination of the potential effects of climate change on the performance of key sources (or groups of sources). A programme of source investigations has been agreed, and is currently being implemented. At this stage, the objective is to derive indicative (rather than definitive) estimates of likely changes in source performance under climate changes of specified type, direction and magnitude. Two case studies are presented here.

1. Weir Wood reservoir

Weir Wood reservoir is a small direct supply reservoir located in the headwaters of the River Medway catchment, near Forest Row, Sussex. Despite having a catchment area of only 26.9 km^2, the present 5500 Ml usable storage capacity is suspected of being undersized; the reservoir has a history of filling regularly and rapidly, and is frequently spilling during times when other reservoirs in the Southern region are well below capacity level.

In early 1990, Southern Science conducted simulation studies to

47

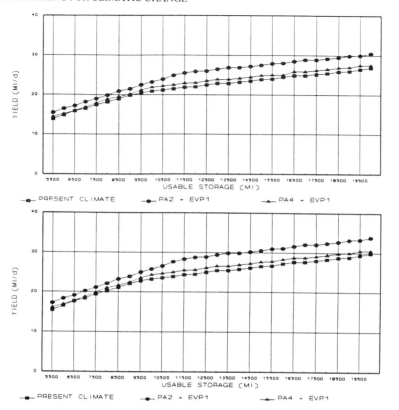

Assumed percentage changes in mean monthly conditions:													
Variable	Scenario	J	F	M	A	M	J	J	A	S	O	N	D
Potential Evapotranspiration	EVP1	40	40	30	25	15	8	8	10	12	20	35	40
Rainfall	PA2	20	20	20	20	10	-10	-15	-10	10	20	20	20
Rainfall	PA4	10	10	10	10	5	-5	-7	-5	5	10	10	10
River flow	EVP1 +PA2	22	19	17	16	11	-4	-10	-8	10	27	24	24
River flow	EVP1 +PA4	8	6	2	1	-1	-6	-7	-5	4	9	6	7
Source: Arnell et al.(1990)													

Fig. 2. Estimates of the reliable average and peak yields obtainable from an enlarged Weir Wood reservoir under present and perturbed climate 2% droughts. The upper plot shows storage against average yield. The lower plot shows storage against peak month yield. Scenario details are shown.

48

detemine the reliable average and peak month yields which may be obtained from an enlarged Weir Wood reservoir (Fenn, 1991). A FORTRAN program was written to determine the yields sustainable from a reservoir of 5500 Ml to 20 000 Ml usable capacity, given 2% drought inflows and losses due to evaporation, prescribed compensation flows, spillages and a seasonally-varying draw-off pattern. The drought inflows were derived by a 'proxy' transfer of the stacked 5 year 2% design drought sequence for the Medway at Teston. This involved creation of a monthly design drought sequence from Weir Wood inflow records corresponding to the months which constitute the Teston design sequence, and derivation of Weir Wood:Teston conversion factors which were used to transfer the 1826 day Teston sequence to Weir Wood. For simplicity, the reservoir was assumed to be full at the start of each simulation, irrespective of reservoir size. The calculated yield-storage curves (Fig. 2) thus continue to climb as storage rises, but may be drawn parallel to the abscissa at whatever upper limit for initial storage is deemed appropriate. During an average year, there is an excess inflow (over evaporation compensation flows and water supply demand) of around 3000 Ml. Thus three average years prior to a drought year provides a refill potential of 9000 Ml (plus any residual capacity), four average years gives 12 000 Ml, and so on.

Simulation using the current climate proxy drought sequence defines a reliable yield for the existing 5500 Ml usable storage of 13.9 Ml/d (cf. an assessed yield of 14.0 Ml/d). Results for different reservoir sizes (Fig. 2) suggest that a reservoir of 10 000 Ml would be optimal in terms of cost and security considerations. Enlargement of the existing reservoir to this capacity would increase average yield by 7 Ml/d. Given rising pressure on resources in the Central Sussex resource area, this is an attractive option. Given the possibility of climate change, it would be imprudent for the company to proceed without determining the likely effects of anticipated climatic changes on the yield of the reservoir. Cole et al. (1991), for example, have suggested that reservoir yields in south-east England may be reduced by 8% to 15% under a best estimate climate change scenario with an implied 8% reduction in annual runoff.

To study the effects of climate change on Weir Wood reservoir specifically, the Weir Wood proxy drought sequence was perturbed to correspond to two different climate change scenarios. The scenarios chosen were the PA2 + EVP1 and PA4 + EVP1 scenarios defined in Arnell et al. 1990); the PA2 + EVP1 scenario defines a condition of 20% additional winter rainfall and 15% less summer rainfall than at present, coupled to an evapotranspiration regime around 15% higher than at present; the PA4 + EVP1 scenario defines winter rainfall increasing by around 10%, summer

49

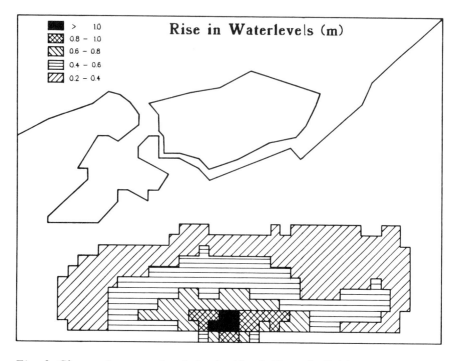

Fig. 3. Change in water levels in the North Kent chalk block associated with a climate change from present conditions to PA2 + EVP1 conditions (the map shows the North Kent coastline from the Isle of Grain to Herne Bay (including the Isle of Sheppey) and the North Downs aquifer between the rivers Medway and Great Stour)

rainfall decreasing by around 7%, and evapotranspiration increasing, as before, by around 15%. Arnell *et al.* (1990) have computed the net effect of these scenarios in terms of the percentage change in mean monthly flow in the Medway at Chafford Weir. On the assumption that these percentage change factors may be used to represent changes in inflow to Weir Wood reservoir during drought periods, the proxy drought sequence was modified to reflect both change scenarios, and the simulation model re-applied to the resulting modified drought inflow sequence. The results obtained are plotted as yield versus storage curves in Fig. 2.

The results show that there would be a yield gain at all storage states were either climate change scenario to occur. Taking the 5500 Ml value to represent the present reservoir, there would be an average yield gain of 1.6 Ml/d (11.5%) under the PA2 +EVP1 scenario, and one of 0.6 Ml/d (4%) under the PA4 + EVP1 scenario; the corresponding peak month yields gains would be 1.77 Ml/d and 0.66 Ml/d. For an enlarged reservoir of 10 000 Ml usable storage, there would be an average yield gain, over the present yield of 13.9 Ml/d, of 7.0 Ml/d (50%) under the unperturbed climate scenario, 9.3 Ml/d (67%) under the PA2 +EVP1 scenario, and 8.0 Ml/d (58%) under the PA4 + EVP1 scenario; the corresponding peak month yields gains would be 7.77 Ml/d, 10.32 Ml/d and 8.88 Ml/d. On the basis of these scenario studies, it would thus seem that climate change may result in net gains in yield at Weir Wood reservoir.

2. North Kent Groundwater Scheme

Groundwater levels, heads and mass balances in the North Downs chalk aquifer between the Rivers Medway and Great Stour, Kent, have been modelled with a two-dimensional finite difference model in order to determine best management practice under current recharge regimes (Southern Water Authority, 1989). The updated model computes a monthly-interval 26 year recharge sequence for each of its 500 nodes using a full Penman-Grindley procedure based on local daily rainfall and monthly evapotranspiration records for the 1960–1986 period. A rapid recharge mechanism is built into the model.

The model has been used to determine the difference in aquifer water balance resulting from a change in recharge regime, from the current regime to one based on a PA4 scenario perturbation of the rainfall sequence and an EVP1 scenario perturbation of the evapotranspiration sequence. (See Fig. 2 for details of the monthly percentage change factors of both scenarios). A repeating cycle of 1989 actual abstractions was used for both simulations — 24 PWS sources (Southern Water Services and Mid-Kent Water) and 10 industrial supply boreholes.

The water balance over the last three years of the 26 year simulation period was used to assess the overall differences resulting from the two recharge regimes. Fig. 3 shows the computed difference in groundwater level. It is clear that equilibrium groundwater levels are higher under the perturbed climate than under the current climate, by up to 1.0 m in places and by over 0.2 m across most of the aquifer (average 0.3 m).

Interim conclusions from scenario studies

Both scenario studies described here suggest an improved resourcesituation under climate change of assumed direction and magnitude. The implication is that in both of these individual cases, the increased winter rainfalls outweigh the increased evapotranspiration and summer rainfall losses to provide increased resources. That this is the case for both a surface water reservoir and a groundwater scheme is hydrologically interesting and operationally encouraging. It would obviously, however, be unwise to extrapolate the results from these two studies to the region as a whole, or to assume that the results are definitive. Reservoir refill and groundwater recharge regimes may be prove to be different under drier spring and autumn scenario conditions (such as those of the UK CCIRG best scenario), and the sensitivity of results to different scenario conditions will need to be carefully examined on a case-by-case basis. The fact that the long term Chilgrove Well groundwater level record (1836–present) contains no historical precedents for the late recharge patterns of recent years (Munnery, 1991) is a salutary caution to complacency in this context.

Given the present state of input scenarios, it is considered that basic surveys of many sources are more valuable than a sophisticated study of a particular source. The likelihood is that results will be sensitive to scenario conditions, and to local circumstances (cf. the results presented here with those of Cole et al. (1991) and Arnell (1992)).

Conclusions

Southern Water Services' initial strategy in respect of climate change has been to develop tools to enable it to determine the consequences of any given change in source drought outputs or demand components, irrespective of whether these arise from climate change or other influences (e.g. domestic metering). Should any change in drought outputs or drought demands now occur, the company is in a position to immediately reckon their consequences in terms of its overall ability to service demands. The same system which provides this capability is now also being used to assess the company's vulnerability to changes in demand, and to changes in source outputs. By 'playing tunes' with the model, the company is able to determine its present safety margins and is able to define the scale of change which would constitute too great a risk to its ability to meet its statutory obligations. The model thus allows the company to determine its sensitivity to climate change.

In tandem with this facility, the programme of perturbed state modelling studies which the company has initiated will enable it to make best possible investment decisions with regard to future resource scheme developments. For the present, 'back of envelope' scenario studies of the type applied to the Weir Wood reservoir and the North Kent Groundwater Scheme problems described herein are being used to assist in the formulation of 'no regrets' forward planning schemes. As confidence in the scenarios grows, these studies will be extended and enhanced to provide better estimates of likely effects resulting from likely changes.

Acknowledgements

The groundwater modelling work was undertaken by T. Keating, Southern Science Ltd.

References

Arnell, N.W. (in press, 1992) Impacts of climatic change on river flow regimes in the UK. Journal of the institution of Water and Environmental Management.

Arnell, N.W., Brown, R.P.C and Reynard, N.S. (1990) Impact of climatic variability and change on river flow regimes in the UK. Institute of Hydrology Report No. 107.

Cole, J.A., Slade, S., Jones, P.D. and Gregory, J.M. (1991) Reliable yields of reservoirs and possible effects of climatic changes. Hydrological Sciences Journal, 36, 6, 579-598.

Department of the Environment (1991) The potential effects of climate change in the United Kingdom. United Kingdom Impacts Review Group First Report, HMS0.

Fenn, C.R. (1990) Climate change due to the enhanced greenhouse effect: implications for the interests of Southern Water Services. Southern Science. Report No. 90/R/114 (Confidential report for Southern Water Services Ltd.).

Fenn, C.R. (1991) Sea level rise: implications for the interests of Southern Water Services. Southern Science Report No. 91/R/150 (Confidential report for Southern Water Services Ltd.).

Fenn, C.R. (1991) Simulation modelling to determine the reliable yield of an enlarged Weir Wood reservoir. Southern Science Report No. 91/R/127 (Confidential report for Southern Water Services (Sussex Division) Ltd.).

Fenn, C.R. (1992) Water Resources Review version 2. Southern Science Report No. 92/7/298 (Confidential report for Southern Water Services Ltd.).

Giles, D.M. (1989) Water Resources Review. Southern Science Report No. 89/R/4 (Confidential report for Southern Water Services Ltd.).

Giles, D.M. (1991) Comments on the calculation of level of service indicators in DG1 and DG4 returns. Southern Science Report No. 91/R/138 (Confidential report for Southern Water Services Ltd.).

Houghton, J.T., Jenkins, G.J. and Ephraums, J.J. (eds.) (1990) Climate Change: the IPCC Scientific Assessment. Cambridge University Press.

Munnery, K.E.F.M. (1991) Groundwater level recessions in the chalk aquifers of SE England. Southern Science Report No. 91/R/151.

(Confidential report for Southern Water Services Ltd.).

National Rivers Authority (Southern Region) (1991) Water Resources Strategy: Draft for Discussion (November 1991).

National Rivers Authority (1992) Water Resources Development Strategy: Discussion Document)March 1992).

Owens, S. (1991) Impact of climate change. Paper presented at the Managing Water Resources in a Dry Decade Conference, Cambridge, 30 September–1 October 1991.

Southern Water Authority (1989) The North Kent Groundwater Development Scheme. Southern Water Authority Internal Report.

Effects of climate change on water resources and demands

C. J. A. BINNIE, MA, DIC, FICE, FIWEM, Director WS Atkins Consultants, and P. R. HERRINGTON, BA, MSc, Senior Lecturer in Economics, University of Leicester

For water engineers and planners it is important to try to understand in quantitative terms the possible effects of climate change on water resources and the demands for public water supplies. In this paper working methods are clarified and indicative magnitudes derived. Broad conclusions are then drawn.

Water resources

Introduction

Any estimates of the impact of future climate change are at best only as reliable as the estimates of the size of the change itself. The best bases we currently have to predict the effect of global warming on our climate, and hence water resources are the Global Circulation Models (ref. 1). Unfortunately the scale of most climate models is about 400 km × 400 km. This means that only one or two grid squares cover the whole of the UK. The biggest problem with using climate model results is that regional predictions of climate change are extremely uncertain. The climate models, therefore, cannot be used directly to predict climate variations across the UK. The approach adopted, therefore, is to use the climate model as the basis for general information to go into hydrological models.

Criteria

The criteria for changes over the next 40 years assumed for this symposium, based on the UK Climate Change Impacts Review Group 'best' rainfall scenario, are set out in Table 1. Precipitation changes in spring and autumn are assumed to be intermediate.

Runoff

The principal factors affecting the mean annual runoff volume are rainfall and potential evapotranspiration. Higher temperatures would increase

Table 1. Climate change scenario

Time Scale	40 years, ie 2032
Global mean surface warming	3 - 4°C
UK mean surface warming Precipitation : average	2-4°C +10% in winter no change in summer
Precipitation : extreme	increased frequency, eg, 1 in 50 years becomes say 1 in 40 years

potential evapotranspiration (PE). However, PE would also be influenced by humidity, windspeed, net radiation and any plant changes, due to higher CO_2 and other effects (ref.2).

Arnell (ref.2) highlights the uncertainty in predictions of both rainfall and PE, and hence runoff, coming from current research predictions of the potential impact of climatic change. Using the 'wettest' (+16% winter, +16% summer) and 'driest' (+0% winter, −16% summer) CCIRG rainfall scenarios and two possible PE scenarios (+7%, and +15% increases in annual PE), he showed that the effect on river runoff across the UK could either increase the annual runoff volume by 30% or decrease it by 30%. He then went on to show that, using the 'best' CCIRG rainfall estimate, the mean annual runoff volume across the UK could either increase by around 4% or decrease by around 8%, depending on which PE scenario is used. Thus, a 7% difference in PE estimates, using the 'best' current estimate of future rainfall, results in either an increase, or a decrease in annual runoff.

The foregoing serves to illustrate two principal points. Firstly, the range of possible future climate scenarios results in such a broad spectrum of future runoff possibilities that specific prediction is not only meaningless, but also misleading. Secondly, estimates of runoff across large areas of the UK are presently very sensitive to estimates of future PE and small changes can result in relatively large impacts on runoff volume. Both of these points highlight the extreme uncertainty in our current forecasts of the possible effects of climate change on river runoff, and really set the scene for this paper. All interpretations on the possible impacts on water resources should be seen against this backdrop. Nevertheless, it is still both necessary and appropriate to consider the possibilities of change and their likely impact on water resources, based on 'best' forecasts, as long as the

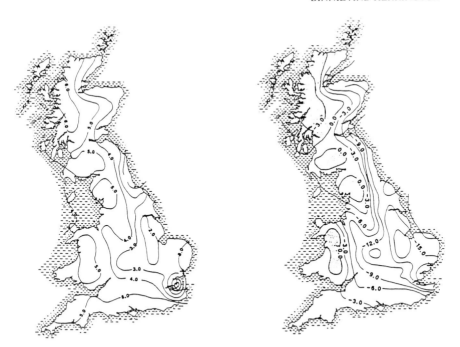

Fig. 1 (left). Percentage change in annual runoff: best CCIRG scenario with potential evapotranspiration increased around 7%.
Fig. 2 (right). Percentage change in annual runoff: best CCIRG scenario with potential evapotranspiration increased around 15%

margin of uncertainty is always kept in mind.

Arnell, Brown and Reynard (ref. 3) have analysed the sensitivity of UK river flow regimes to climate change using a simple monthly water balance model. The model used a continuous accounting procedure to portion rainfall between evapotranspiration, replenishment of soil moisture storage and river runoff using parameters representative of an 'average' UK catchment. It was applied to a disturbed series of 40 × 40 km gridded monthly climate data from the MORECS archive (ref. 4) to produce maps of possible change in runoff across the country under different scenarios

Table 2. Percentage changes in potential evapotranspiration over current averages

J	F	M	A	M	J	J	A	S	O	N	D	Anl
20	20	15	12	7	4	4	5	6	10	17	20	−7

(Figs 1 and 2).

Arnell (ref. 3) took two alternative scenarios of evapotranspiration and the CCIRG 'best' estimate of rainfall (+8% winter, 0% summer) to illustrate the range of likely effects. The assessment in this paper is developed from Arnell's analysis.

Scenario 1

The first scenario, EVP2, is set out in Table 2. The resulting percentage runoff is shown in map form in Fig. 1.

This scenario is only marginally different from the change which has been suggested for the conference of an increase in PE of 5 per cent across the UK as a whole. However there are significant variations in runoff across the country. Arnell (ref. 2) has classified catchments into three general groups: upland, lowland, and groundwater-fed. This paper considers the possible impact of climatic change on water resources in each of these three types of catchment.

Upland catchment

Upland catchment rainfall at present closely matches PE during summer. Any change in climate could tip the catchment into surplus or deficit and thus has a relatively large impact on the water available for river runoff. Thus, under EVP2 assumptions, summer runoff of a particular but typical catchment could drop by 6%, with increases in the rest of the year of 6% – 8%.

Most upland water sources are reservoirs. These were often designed to provide a yield under conditions of a 1 in 50 year drought. Many of these reservoirs reach their critical design condition of minimum storage level at the end of one summer season.

Under this scenario there could be less problem of reservoir refilling due to increased winter runoff. Small changes in spring and autumn conditions are likely to result in little change in the actual start and finish of the critical drawdown period. Reduced summer inflow, possibly around −6%, would reduce the water available for supply during this critical period.

Larger upland reservoirs may have storage volumes as much as 30% of the mean annual runoff. For these the reduction in water available could be about 1%. However, there are many upland reservoirs with storage volumes of smaller percentages of annual runoff. For these, annual draw-off reductions could reach several per cent. Each would need individual analysis.

Lowland catchment

Using the 'best' CCIRG rainall scenarios, the analysis showed that summer runoff in a particular lowland river unchanged with winter flows increasing by about 9%.

Under this scenario the critical drawdown period would generally be similar to that at present. For smaller lowland reservoirs which are single season critical there would be little change in water available during drawdown.

Some large lowland reservoirs are double season reservoirs, i.e. they do not fill in the middle winter of the critical period and maximum drawdown is reached at the end of the second summer. Thus the critical period covers two summers and one winter. For these reservoirs under this scenario, the postulated increase in winter runoff would be expected to increase the overall yield.

On lowland rivers there are a number of run-of-river direct abstraction schemes, unsupported by storage. Most of these abstractions now have minimum flow conditions below which abstraction may no longer take place. The effect on water available would depend on the relationship between river flow, minimum flow and amount taken and could vary from no effect to considerable effect.

Many lowland reservoirs are, in fact, pumped storage schemes, pumping flows from a large river into a reservoir on a minor tributary such as Grafham and Rutland reservoirs, or into a totally embanked reservoir such as those alongside the Thames. For these the impact would vary with the river flow, the prescribed flow conditions in the river, and the volume of

Table 3. Percentage changes in potential evapotranspiration over current averages (EVP1)

J	F	M	A	M	J	J	A	S	O	N	D	An1
40	40	30	25	15	8	8	10	12	20	35	10	-15

Table 4. Comparison of conference scenario percentage change in seasonal runoff of upland catchments with EVP1

	Winter DJF	Spring MAM	Summer JJA	Autumn SON
EVP1	+7	-4	-13	+4
Conference	+5	+5	-15	+5

Table 5. Percentage change in seasonal runoff of lowland catchments with EVP1

	Winter (DJF)	Spring (MAM)	Summer (JJA)	Autumn (SON)
EVP1	-2	-6	-4	0
Conference	+8		-4	

storage. Each case would need considering on its own merits and no general guidance could be given.

Scenario 2: Evapotranspiration

Let us now look at an alternative evapotranspiration PE scenario, EVP1, as set out in Table 3. This is not dissimilar to the conference criteria of summer potential evaporation increasing by 10%. Winter temperatures are expected to rise more than the summer temperatures (ref. 5) so it is likely that the low winter potential evaporation could increase much more in percentage terms than the summer, although in quantitative terms, the winter increase is not likely to have much impact. Again, the analysis is based on the CCIRG 'best' rainfall scenario.

For comparative purposes it is interesting to note that PE during 1989 was 15% higher than the long term average across much of southern and eastern Britain (ref. 6).

Upland catchment

For a typical upland catchment the comparison between Arnell's analysis and the conference runoff scenario is set out in Table 4.

The two runoff scenarios compare reasonably well, although there is a discrepancy in estimates of spring runoff. The results of Arnell's analysis of percentage change in annual runoff across the country are shown in Fig. 2. This indicates that, under scenario 2, upland areas are confined to Wales, parts of the Pennines and Lake District, and Western Scotland. Using the same analysis as in the previous section large single season reservoirs could experience a drop in average draw-off by a few per cent. Water available in an average year from smaller reservoirs could drop significantly more.

Lowland catchment

For a particular lowland catchment the comparison of the runoff

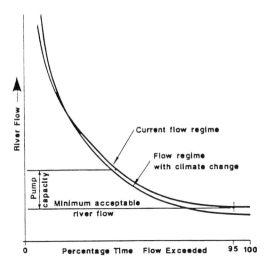

Fig. 3. Possible effect of climate change on flow duration curve for a pumped storage scheme in lowland England

scenarios is set out in Table 5.

Figure 2 shows that there are large areas of eastern England where, for catchments not supported by groundwater or treated effluent return, average annual runoff could drop by over 10%. Summer runoff could decrease appreciably more.

This scenario compares with the United Kingdom Statement to the UN Conference on Water and the Environment 1992 which stated that under these climatic conditions "model analyses showed that average annual flows in parts of south east England would fall by over 20 per cent" (ref. 7).

Many lowland reservoirs are pumped storage schemes abstracting from a nearby river. Generally, minimum flow conditions are set in the river below which abstraction may not take place. In the past these have sometimes been set at about the level normally exceeded 95% of the time. In the south and east, under Scenario 2 (Fig. 2), most flows would reduce, as shown indicatively in Fig. 3. This shows that the pumps would run for shorter periods of time and hence would pump less water into the reservoir. Most of any extra runoff in the wetter winter periods would occur when the pumps were already operating at full capacity and hence the water would be 'lost'. Thus, pumped storage schemes in the south and east could be affected more than reservoirs. In this case double season pumped reservoirs, such as Graffham, could be worse off, because little extra water would be obtained in the intermediate winter but appreciably less in the

61

rest of the year. Each needs individual assessment against the latest runoff predictions and the actual scheme configuration.

Yield analysis

The analyses in this paper so far have considered average years. Climate change will also affect variability. It is generally believed that extremes will occur more frequently. The conference scenario suggests that a frequency of 1 in 50 years will become 1 in 40 years. Thus reservoir yields for fixed return periods will drop more than the average water available quoted above.

Cole *et al.* (ref. 8) have prepared generalised graphs of yield change based on certain assumptions and the analysis of certain reservoirs. Their findings were that, for the criteria they selected and single season impounding reservoirs with storage over 1/4 of the mean annual runoff, yields generally dropped by about 10% in the South East (lowlands) and, in the North West, (upland), by about 5%.

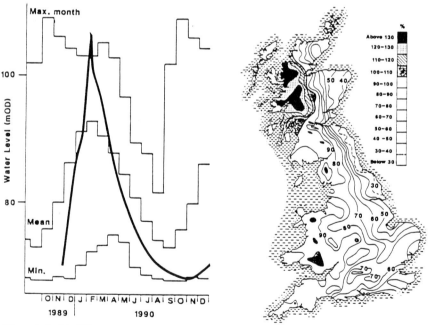

Fig. 4 (left). West Woodgates Manor observation borehole
Fig. 5 (right). A guide to 1989 runoff expressed as a percentage of the 1961 – 1988 average

Groundwater

Most groundwater sources are in lowland (SE) England. As a general rule groundwater recharge in Britain takes place in winter once the summer soil moisture deficits have been replenished, and ceases in late spring when soil moisture deficits reappear.

In considering the effects of climate change on groundwater recharge there are several conflicting factors. Greater winter rainfall should result in greater recharge. Greater summer evapotranspiration would, however, absorb more of the autumn rainfall and is likely to mean a later start to the recharge period. Any increase in winter evapotranspiration would also diminish the benefits of the extra winter rainfall.

Almost all the groundwater areas are located in the lowland areas of the country. Thus the winter runoff under the two alternative evapotranspiration scenarios and the best CCIRG rainfall scenario varies from +9% to −2%. However, under average conditions it is possible that the shorter recharge period would balance the extra winter water available and total recharge would change little (ref. 7).

In some areas any recharge increase may be limited by the constraints of aquifer properties, e.g. in the Permo-Triassic sandstones which have lower infiltration capacities than the Chalk and the Upper Greensand aquifers (ref .9).

In the Chalk and Greensand in the south-east there is also some evidence from the very wet but short recharge in early 1990 that shorter wetter recharge periods fill the fissures but do not have time to permeate the full aquifer. Thus water levels near the all-time high at the end of the 1990 recharge season were approaching all-time lows by the end of the summer, see Fig. 4.

An additional parameter is the prospect of altered infiltration characteristics of some soils modified by drought or by more humid winters.

General comments

Many of the assessments in this paper are tentative. Thus criteria and predictions do and will change. However, a number of comments can be made.

(a) Figure 2 indicates a considerable spatial variation in change in runoff across the country.

(b) The Climate Change Impacts Review Group (ref. 10) reported that, in the south-east of England in particular, water resources are highly sensitive to small changes in rainfall and evaporation. This can be seen by comparing Fig. 1 with the alternative scenario in Fig. 2.

(c) It is not possible to predict that the outturn will be closer to one

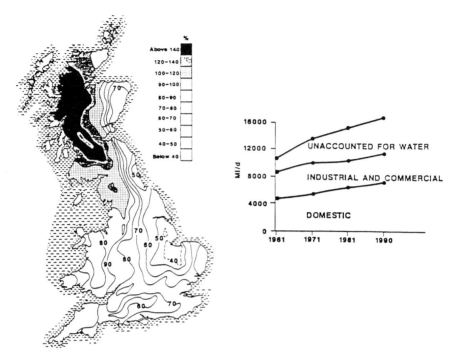

Fig. 6 (left). A guide to 1990 runoff expressed as a percentage of the 1961 – 1989 average

Fig. 7 (right). Potable public water supplied in England and Wales 1961 – 1990 (Ml/day)

scenario than another, and thus there is considerable uncertainty in present predictions.

(d) Climate change is, however, likely to accentuate the current wetter north and west and drier south and east situation.

It is interesting to see how the runoff patterns have changed recently. Fig. 5 shows the situation in 1989. We would stress that this pattern is not actual runoff but a percentage of the long-term average. This shows a dramatic difference, with Wales and West Scotland becoming wetter and the remainder of the country drier, particularly the east coast. One year's change does not indicate climate change, but compare this with 1990, Fig. 6. There is a very similar pattern.

We believe these climate conditions were caused by high pressure systems sitting over Northern France much more often, forcing the depressions and weather fronts north-westwards to the west coast of Scotland.

It is also interesting to compare recent percentage runoff, Fig. 6, with one of Arnell's scenarios of climate change, Fig. 2. The spatial similarity is remarkable. It is of interest that when the 1881–1915 averages were replaced by the 1916–1950 averages, significant regional differences appeared with an increase of 14% in the west of Scotland, and in some places a decrease on the eastern side of UK (ref. 11).

Given these uncertainties, what are the implications for water resources management? It is likely that both the seasonal and regional distribution of resources will change, with an increasing concentration of available water in winter and in northern and western catchments (ref. 12). This means we could no longer rely totally on historical records to predict the variability of river flow. Supplies during the warmer, possibly drier summers would need to be maintained by larger storages or inter-basin transfers.

Several authors have suggested that water resources engineers should adopt a flexible approach. What should water resources engineers do? Should they oversize new draw-off works in case they can get more yield from reservoirs? Where yields may reduce, should they design dams so that they can be raised later to provide the extra storage needed to compensate? The latter requires strengthened draw-off tunnels etc. and the positioning of other structures away from the toe of the dam. This would cost money. In the days of a privatised water industry it would be difficult to justify such extra costs now without stronger evidence of future change.

Water engineers do have one help and that is that these effects will build up over a period of time, currently estimated at 40 years. However, there are no recent new reservoirs completed in much under 15 years from inception.

Thus it will be appropriate to adopt a flexible approach to water resource development, to follow the latest research predictions and to carry out sensitivity analyses of alternative scenarios when appropriate.

Resources: conclusions

More research needs to be done on the likely effects of climate change on the hydrological cycle, particularly refining confidence in predictions of rainfall, potential evapotranspiration and resulting change in runoff.

If the changes postulated under current best estimates occur, then most reservoirs, sources may not be affected much, generally less than 10%, many less than 5%.

When planning new sources each source needs to be studied

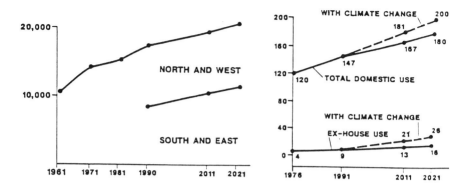

Fig. 8 (left). Public water supply abstractions and NRA 'baseline' forecasts (Ml/day)

Fig. 9 (right). Per capita domestic consumption: component forecasts to 2021 for the SE (lhd)

individually using sensitivity analyses.

When planning water resources development, we need to adopt a flexible approach, but the justification of extra expenditure may need stronger evidence than is currently available.

Water demands

Introduction

All demographic and economic forecasts in the UK predict smaller households, population increases in southern and eastern regions, increasing living standards and more leisure time. The demands for water services at present real prices and with unchanged use regulations would thus be likely to continue to increase significantly into the longer term future (i.e. 20 to 40 years ahead). Climate change will affect this prospect, increasing both estimated demands and the uncertainty that surrounds them. What are the possible magnitudes involved?

This section of the paper is restricted to public water supplies. It first maps out the past and the water companies' own expected future demands in the absence of climatic change, and then discusses the sensitivity of forecasts to a 2°C warming over the next 40 years. Provisional estimates of orders of magnitude will be presented for both average and peak demands.

Current demand projections

In recent decades average per capita domestic demands for piped water in England and Wales have been increasing at about 1½ litres per head per day each year. Combined with a small (10%) net increase in metered consumption in the last 30 years and a growing unaccounted for water (UFW) element, this has meant that total public water supplies have risen by about 60% since 1961.

Recently the NRA has assembled and published current water company demand forecasts (refs 12, 13). These suggest an increase in average daily demands of 18% in the 30 years to 2021 (see Fig. 8). The slower growth rate is presumably largely due to substantial reductions in UFW as mains replacement and leakage detection programmes take effect. Individual water companies' sectoral forecasts must apparently remain secret, although it is known that the 1990–2011 forecasts reflect per capita domestic demand increases of up to 40 l/hd (NRA, personal communication).

What is very clear from the NRA documents, however, is the emergence of a dramatic two-nations picture of future water projections. In 1990 both the population of England and Wales and aggregate PWS demands divided equally between the five NRA regions in the south and east (SE) and the five in the north and west (NW). In the next 30 years (see Fig. 8), the water companies' — and thus the NRA's — baseline prediction is of an increase of 33% in the SE and only 4% in the NW. If no additional resources were developed, such demands would give a reliable yield shortfall of 18% in the south-east but a surplus in all other

Table 6. SE Water Company peak week ratios 1990 onwards

1990 range, reported to NRA 2011 forecast range, reported to NRA	1.09 - 1.66 1.16 - 1.73

1990 - 2011 forecast
increases*, reported to NRA
  10 companies in range
  7 companies in range
  4 companies in range
	0.00 - 0.05
	0.05-0.10
	> 0.15

+4°C effect, from econometric equations	+0.04 to +0.07
+4°C effect, from components forecasting	+0.07 to +0.10

Sources: (Ref.12); results of Leicester University research.
Note: * believed to be in absence of climate change

Table 7. Feasible present and future domestic and PWS 7-day peak ratios

	1991		2021 No Climate Change		2021 With Climate Change	
	Avg	P7D	Avg	P7D	Avg	P7D
In House (1hd)	137	151	164	180	174	191
Ex House (1hd)	10	47-55	16	75-88	26	122-143
Domestic (1hd)	147	198-206	180	255-268	200	313-334
Domestic P7DRs	1.35-1.40		1.42-1.49		1.57-1.67	
PWS P7DRs	1.17-1.24		1.21-1.31		1.28-1.41	
Effect of Climate Change on PWS P7DR					+0.07 to +0.10	

Sources: from Table 8 and assumptions described in text.

regions. The implication of such forecasts must be a very substantial increase in average per capita domestic consumption in the SE, from 147 to about 205 l/hd, and much smaller growth in the NW, from 133 to about 165 l/hd.

Climate change analysis

That is the backcloth of 'official' forecast demands, with all the uncertainties they must contain. Now we mix in climate change. Three ways into the problem suggest themselves, and all these are being pursued in work at present being undertaken for the Department of the Environment by Leicester University's Department of Economics: (i) econometric modelling of average demands and peak ratios; (ii) prediction of future domestic use by consideration of future ownership, frequency of use and 'volume per use' for individual household water-using appliances and habits; and (iii) international comparisons, paying particular attention to the experience of warmer developed economies in respect of these elements of water use thought to be most sensitive to climate. For a general introduction to water demand forecasting, the reader is referred to Gardiner and Herrington (ref.14).

Econometric modelling

In the first method attempts are made to 'explain', using regression analysis, past variation in sectoral or aggregate PWS time series and in various monthly/weekly/daily PWS ratios (peak and otherwise), with the aid of appropriate economic series (e.g., income, wealth, production) and climatic variables. Predicted or imposed future values of the 'independent' variables may then be inserted into the derived equations and forecasts thus generated, with confidence limits if required.

Because so few consistent time-series for domestic consumption are available in the UK (due to the absence of domestic metering) most work investigating average use and ratio variations has had to concentrate on unmeasured and aggregate PWS respectively. Past experience and the provisional results of present investigations may be summarised as follows:

- In England and Wales as a whole a hot and dry restrictions-free year like 1975 is associated with water demands 1½% to 2% higher than those of a climatically average year.
- In May (and, recently, April) the most important climatic variable is rainfall, explaining 25% to 65% of variations in some monthly PWS ratios in the SE.
- Over the June – August period, peak seven-day (P7D) and weekly/monthly ratios show closest correlation with average maximum daily temperature. Overall, climate can explain between 30% and 85% of the year-to-year variation in ratios.
- The influence of temperature on June – August peak and non-peak ratio series is found to be surprisingly stable. Analysing the records of six divisions or companies in the south-east which at present deliver between 25 Ml/d and 2000 Ml/d (and in aggregate are responsible for more than 30% of PWS in the SE), we have found a 1°C rise in average daily maximum temperature increases the ratios by between 0.010 and 0.017. Thus in one company's coastal divisions a 'normal' P7D ratio of 1.30 typically increases to 1.33 if temperature increases from one year to the next by 2°C; and for a large water company division we found the June PWS would typically be running at 8% above the winter average rather than 6% for the same 2°C temperature change.
- Contrary to expectations there is as yet little evidence that summer peak and average use ratios increased in the 1980s, even after allowance is made for climatic variation and supply restrictions. This remains a puzzle.

Simplistic insertion of +2 to +4°C temperature increases in the derived equations would generate ratio increases in the range +0.02 to +0.07. These changes appear small, but they reflect the experience of climate variation rather than change and thus do not incorporate all the changes in water use habits that might be expected to follow permanent warming.

69

*Table 8. Suggested component forecasts for average domestic demands
(unmetered) in non-metropolitan south and east of England* (SE)
(litres/head/day; bracketed figures incorporate climate change effects)*

	1991°	2011	2021
WC Use	35	32	31
Personal Washing	36	44 (50)	48 (58)
Clothes Washing	21	22	23
Dish Washing	12	11	11
Waste Disposal Unit	1	2	3
Car Washing	2	2	3
Lawn Sprinkling	1	3 (6)	4 (9)
Other Garden Use	6	8 (13)	9 (14)
Identified Demand	114	124 (138)	132 (152)
Misc Use *	33	43	48
Total Domestic Demand	147	167 (181)	180 (200)

Sources: based on published (eg, Refs 15-17) and unpublished
studies and author's own estimates.
Notes: * also allows for unforseen developments ° 1991 data
reflect climatically average average year

Note also that the increases are not insignificant when viewed alongside
the water companies' own peak week demand forecast increases reported
recently by NRA; see Table 6.

Components forecasting

Permanent structural changes in demands upon the PWS due to
long-term climate change are best predicted with the aid of components
forecasting, both without and within the domestic sector. Detailed consideration of the traditionally metered sector means breaking it down into
industrial groupings and undertaking forecasts for the output and water
efficiency prospects of each group before re-assembling (see Archibald
(ref. 15) for a good example). Special attention clearly needs to be paid
to average and peak PWS demands by agriculture, horticulture and
food processing, although at present these appear to be of only local
significance.

Air conditioning is also of potential importance, but it appears that for
both permanent plumbed-in and portable systems water-linked technologies may be in decline, because of higher installation costs and perceived

health risks respectively.

Within the domestic sector, components forecasting should be used to examine the possible shape of the longer term future. Fig. 9 presents the results of an exercise undertaken for an imaginary but hopefully typical large non-metropolitan area in the south-east (for full data see Table 8). It makes use of numerous studies concerning actual and predicted appliance ownership, frequency of use, and past and expected future water use technology (see e.g. Archibald, 1983; National Water Council, 1982; and Hall, Hooper and Postle (refs 15 – 17) and also describes the possible effects of climate change by embodying specific plausible assumptions about personal showering, lawn sprinkling and other garden watering (see Table 9).

Present external to-the-house domestic use in the south-east area is estimated as 9 l/hd in a climatically average year, 6% of domestic per capita consumption. Our component forecasts suggest this may increase to 9% of the total by 2021 even in the absence of climatic change. On the assumptions adopted climate change increases the proportion to 11%, although as much demand is added to the in-house total (personal showering + 5 l/hd) as garden watering and lawn sprinkling add outside the house. Overall, our best estimate at present is that climate change adds 4% to average demands in 2011 and 6% by 2021. But it should be emphasised that this is a provisional result; amendments are to be expected as more information is collected.

Peak demands

Climate change can also influence peak demands, normally measured by the ratio of the seven days with the highest demand to the average demand. In Table 7 we calculate 'working hypothesis' peak seven-day ratios (P7DRs) for the data set out in Table 8 incorporated in Fig. 9. In-house P7DR is constrained to equal 1.1 (broadly consistent with diary studies of domestic use undertaken by water authorities in the mid-1970s) and a range for the overall domestic P7DR of 1.35 to 1.40 has been assumed for 1991 in the south-east. The implied 1991 range for ex-house P7DR is thus 4.7 to 5.5, and the 1.1 and 4.7 – 5.5 ratios are then applied to future average demands to generate possible future domestic P7D scenarios. A range of assumptions for the structure of future PWS have been adopted (50% – 55% domestic), the P7DR of non-domestic use assumed to lie in the range 1.00 – 1.10 during the domestic seven-day peak.

On the assumptions made, it is seen that the PWS peak ratio increases by between 0.04 and 0.06 as a result of climate change. These figures have been added to Table 6, where they are seen to be slightly larger than the econometrically derived estimates described earlier.

International comparisons

Work is still in progress on the collection of data from abroad that may enable relevant international comparisons on climate sensitive domestic water use to be made. The problems in this sort of exercise are severe however. Not only does the sort of data required appear to be very scarce, but also differences between countries in 'water use culture' may stand irrespective of climate differences.

Demands: conclusions

Both average and peak PWS demands are likely to grow significantly even in the absence of global warming. Climate change probably means more of the same, especially in the south-east. Very provisional estimates are that average demands in 2021 may be 6% higher as a result of increased temperatures, with peak seven day ratios 3 to 5 per cent higher. Demand increases can in principle be met by supply expansion or demand management (including domestic metering and tariff design).

The right balance should be determined by technical, economic and environmental criteria. But we have shown above that garden watering, which is known to be more sensitive to water price than other domestic use, may be responsible for as much as 20% (no climate change) or 26% (with climate change) of the projected increases in per capita domestic demands over 1991–2021. The case for domestic metering in the south and east will therefore in any case grow more powerful over time; it will become stronger—perhaps irresistible—if global warming does in fact take place.

Overall conclusion

Climate change could augment the demand for public water supplies in the south and east of England by about 6% by 2021 as well as increasing the peak ratio.

The effect on water resources will depend on the actual rainfall and evapotranspiration but, overall, runoff is likely to become more seasonal and variable. Each source would need analysis individually. On current scenarios single season reservoir yields are more likely to reduce but many less than 5%.

Groundwater sources could increase but may not change much.

Table 9. Three climate-sensitive components of domestic demand

---

Personal washing

Climate change assumed to increase 2011 shower use from 80% to 85% of the population and to increase showering frequency from 3 to 4/week. Half the showers are assumed to use 30 l/shower and half are power showers using 60 l. By 2021 95% (rather than 90%) assumed to be using showers due to climate change, with frequency of use up from 3½ to 5/week. Half the stock are then power showers (1991 assumption: 55% shower ownership, 5% of which are power showers; frequency of use = 2.5/week).

Lawnsprinkling

In 2011 sprinkler ownership assumed to be increased by climate change from 40% to 50% and use is assumed to be once every three days in the May – August period (250 l/use) rather than once every five days. In 2021, ownership (and use) unchanged but preferred frequency of use every other day instead of once every three days in May – August (1991 assumption: 20% sprinkler ownership and use, with average of 20 uses per year at 250 l/use).

Other garden watering

In 2011 70% of households assumed to water gardens with hoses, buckets or cans. Climate change assumed to mean daily watering over April – August, rather than 5 times/week. 70 l per 'water' assumed. In 2021 80% of households water, with frequencies and volume as in 2011 (1991 assumption: 70% of households use 100 l per 'water' and water 3.5 times per week.

Because the effect of climate change is not yet well known, a flexible approach is needed. Demand management, including metering, may become more appropriate than supply management with its large costs for new or enlarged schemes.

Acknowledgements

The paper includes some provisional results arising from work undertaken at the University of Leicester and funded by the Department of Environment through an NERC research contract. Opinions expressed are, however, the responsibility of the authors alone.

References

1. Houghton J, Jenkins G and Ephraums J, Climate Change; the IPCC Scientific Assessment Cambridge University Press: Cambridge 1990.
2. Arnell N, Impacts of Climate Change on River Flow Regimes in the UK, IWEM 1992.
3. Arnell N, Brown R and Reynard N, Impact of Climatic Variability and Change on River Flow Regimes in the UK. Institute of Hydrology Report 107, Wallingford, Oxon, 1990.
4. Arnell N, Climate Change and River Flow Regimes in the United Kingdom. Paper to the British Hydrological Society National Meeting December 1991.
5. Wigley T, Santer B, Schlesinger M and Mitchell J, 1991. Developing time-dependent climate change scenarios for equilibrium GCM results. Climate Change (in press).
6. Reynard N, Arnell N, Marsh T and Bryant S, Hydrological Characteristics of summer 1989 and winter 1989/90, Inst. of Terrestrial Ecology, 1990.
7. The United Kingdom Statement to the United Nations Conference on Water and the Environment 1992.
8. Cole J, Slade S, Jones P, Gregory J. Reliable yield of reservoirs and possible effects of climatic change, Hydrological Sciences Journal 12/1991.
9. Cook R, Water Resources, Effects of Climate Change, IWEM meeting 1991.
10. Climate Change Impacts Review Group the Potential Effects of Climate Change in the UK DoE January 1991.
11. Masley G, Effect of Long Term Trends in Weather on Design. Proc: ICE Part 1, 1978 Informal Discussion
12. National Rivers Authority, Demands and Resources of Water Undertakers in England and Wales. 1991.
13. National Rivers Authority, Water Resources Development Strategy Discussion Document. Bristol 1992.
14. Gardiner V and Herrington P (eds). Water Demand Forecasting. Geo Books, Norwich, 1986.
15. Archibald G, Forecasting Water Demand — A Desegregated Approach, Journal of Forecasting, 1983, 2, 181–192.
16. National Water Council, Components of Household Water Demand. NWC Occasional Technical Paper No.6. London, 1982.
17. Hall M, Hooper B, and Postle S, Domestic Per Capita Water Consumption in South West England, Jrnl IWEM, 1988, vol. 2, December, 626–631.

Discussion

Raporteur: P.J. EWING

A contributor asked for some indication of the extra investment required to meet predicted climate change scenarios and whether this had been compared to the costs of rectifying problems as they became apparent.

Mr Fenn and Mr Hewett replied that the necessary investment had not been calculated using this type of comparison, however the additional costs of accounting for climate change would be best assessed when individual projects were completed.

In referring to the strategic transfer scheme, outlined by Dr Swinnerton in the first presentation, Mr Child of the NRA asked if the needs of agriculture, industry and the environment and not just the needs of the domestic consumer had been considered in the plans.

Dr Swinnerton confirmed that these areas were being looked at but that it was currently very difficult to account for the volumes of water that would be involved.

The final question that was relevant to the presentations was put to Mr Herrington. He was asked, by Mr Meakin of SERC, if the figures for the projected increase in water demand took account of any water recycling.

Mr Herrington explained that the figures did not allow for any recycling other than what might expected to result from metered consumers cutting their consumption after receiving large bills.

Energy

I. W. HANNAH, Chairman, Energy Board, Institution of Civil
Engineers

Introduction

The ICE Energy Board has a responsibility to Council for maintaining the
Learned Society activities of the Institution in the energy field. We are a
group of engineering practitioners chosen from a broad band of energy
related civil engineering activity areas. Our aim is to establish a balanced,
overall appreciation of the energy scene on the Institution's behalf, and,
through papers and conferences, to make that knowledge available to our
members. Accordingly my observations reflect our present limited understanding of the effects of climatic change on that energy scene and do not
aspire to being either comprehensive or erudite.

The Energy industries are more the villains of the climatic change
probabilities than the victims. Energy conversion processes in power
production and transport have played a major role in the intensification of
the greenhouse gases in our atmosphere, which in turn has promoted the
climatic uncertainties. There is no doubt that this intensification activity
will continue and worsen over the next few decades. Despite the high
profile international endeavours in the EC and recently of world leaders
meeting in Rio de Janiero, our developed world is not yet sufficiently
concerned to take really radical steps to reduce the production of greenhouse gases by addressing either power or transport. Their present modest
aims of holding the 1990 greenhouse gas production levels for about a
decade, with a more than sporting chance of missing that uninspiringly
low target, hardly points to an international political climate sensing
imminent disaster.

Not unreasonably, the undeveloped world has aspirations of moving
towards the living standards of the first world, with its inherent high energy
conversion needs; our 25% of Earth's population who use 80% of its
current energy resources. Those associated problems were well articulated
by the King of Spain, in opening the 15th World Energy Congress, when
he said, "Quality of life depends to a great extent on the availability of
energy. In our societies energy is a fundamental element. Increasing
technological innovation and new socio-economic structures are necessary for people to obtain, under acceptable conditions, all the energy
resources they need."

Engineering for climatic change. Thomas Telford, London, 1993

The consequences of those third world aspirations on the total levels of greenhouse gas production will obviously be dependent on the rate of their achievement, but it takes no imagination to appreciate the scale of increase in greenhouse gas emissions as the undeveloped world's billions move, however slowly, towards the profligate American energy consumption of some 12 000 tonnes of coal equivalent per person. These gross figures graphically illustrate the potential problem.

It is worth noting here that the World Energy Congress was attended by some 3000 expert delegates, who devoted much of their work to this problem, exchanging innumerable fine words of principle. No paraphrasing of those well intentioned deliberations is possible in this paper, but there is perhaps some comfort to be drawn from the fact that this critical issue is now becoming a standard item on the meeting agendas of our planet's great and good.

What effective, practical measures will evolve from such high level techno-political debates, and how soon, remain the real issues. There are perhaps six non-contentious, fundamental areas for advance that offer real hope of exercising some control on the burgeoning production of greenhouse gases, when considering the longer term future:

(1) Improving existing conversion efficiencies, especially in the third world.

(2) Developing improved conversion technologies, both technically and environmentally.

(3) Pursuing energy conservation vigorously.

(4) Accelerating R&D on the renewable generation technologies.

(5) Improving both the technology and the public acceptability of nuclear power.

(6) Improving the efficiencies of energy using devices and transport.

In the short term however, it may be argued that, faced with such a massive worldwide potential increase in greenhouse gas production, there is nothing much that we in the UK can possibly do that will influence the outcome significantly. Even stopping all power and transport fossil fuel burning tomorrow, were that remotely possible, would have a negligible world effect. Although I am sure that the developed world would not wish

77

to take any such irresponsible, hopeless and helpless attitude, it is important to realise how limited is the UK's influence on this problem. In this matter we are indeed a small offshore island!

Let us now consider the effects of the changes given in our working scenario on the UK power industries — the power industry as victim. I begin by reminding you of the key figures given in the basic scenario paper.

Changes over the next 40 years

Temperature: 1 – 4°C rise.

Precipitation: 10% increase in winter.

Storminess: + or – 5% in high winds.

Sea level: 14 – 24 cm rise.

Low level inversions: probable increase.

As an initial generality, it is possible to say that the effect of such variations on the UK energy scene are unlikely to be radical. Even if these changes were to occur suddenly, I have no doubt that the diversity of supply existing in the power production industry would smooth out the resulting problems fairly easily. In reality, these slow transitions occurring over some 40 years are most unlikely to throw up any critical problems for the industry. It is significant here to remember that a complete construction-generation-decommissioning cycle for either a fossil-fired, a nuclear station or even a CCGT is a similar time span of about 40 – 50 years. Hence there will always be scope for "designing out" any engineering problems induced by such climatic changes as they arise, during the continuous rolling programme of replacement power stations, thereby leaving the newer, more efficient stations always designed to meet the actual climatic conditions they are likely to encounter. As the equivalent "construction and use" cycles in the transport industry are considerably shorter than for power plant, this "designing out" option is even more favourable for accommodating any necessary transport changes with time.

I trust that this generalisation does not sound too complacent. In fact there will be significant problems and perks induced by such climatic changes. Existing power stations may well require some consequential capital works modifications through their lives. They are even more likely to need variations to their operational regimes to accommodate climatic change. However, I see no potential need for radical reactions in the near future. This range of climatic change will introduce no call for "show stopping" measures. Both the power and transport industries will progressively need to up-date their design requirements rather than to indulge in "back to the drawing board" dramas.

Nevertheless, there will be many less dramatic consequences in the energy field if our scenario of change does occur. I have chosen four for illustrative purposes, but recognise that several others could be regarded as being of at least equal importance. My selection is as follows:

(1) changing demand patterns,
(2) cooling water systems — performance and design,
(3) siting and economics of tidal barrages and wave power devices,
(4) hydro-electric stations with increased precipitation.

Changing demand patterns

As has already been pointed out, the seemingly small climatic changes in our scenario figures are, in approximate terms, equivalent to imposing Mediterranean weather conditions on to the UK. Relating this change to known domestic power demand patterns in the two regions strongly suggests that the UK winter heating peaks will be ameliorated and the summer air-conditioning demand enhanced. Industrial patterns of demand will probably move similarly but less significantly. The effect of such changes on the entire UK system will probably cause a reduction in the overall peak demand figures, with the possibility of that peak eventually occurring in summer as air conditioning gains popular favour. In plant terms this suggests a mild reduction in capacity needs, which may well be scarcely detectable amongst the several other relevant variables. It equally could introduce a problem of re-phasing major plant maintenance outages, as the overall average demand is evened out over the year, filling in the summer demand trough which presently accommodates this essential activity

Cooling water systems
Performance and design

Our scenario's climatic changes would undoubtedly influence power station cooling system performance and design. Directly cooled systems associated with coastal stations are carefully sited relative to mean tide levels in order to minimise pumping costs. Even such apparently small mean tidal height changes as our $14 - 24$ cm can readily be seen to represent significant additional pumping costs, when it is recalled that the closed loop cooling system may well be circulating some 130 000 m^3 of water per hour for every 1000 MW of station output. Furthermore, the raised ambient temperatures would directly detract from the station condenser's cooling performance to a noteworthy degree. This in turn would lower the overall station thermal efficiency, thereby requiring more fuel to be consumed to maintain the electrical output and, of course, simultaneously adding to the emissions of greenhouse gases.

Our directly cooled coastal systems would also have an enhanced risk of flooding. On all stations, but especially on nuclear ones, the level of flood defences around the cooling system is of considerable importance. Examples of pump house inundation in remarkably short periods have been experienced, due to unfortunate coincidences of highest astronomical tide conditions with abnormally high wind driven surges. Sea defence reviews would therefore become a necessary on-going calculation for coastal stations if our predicted tide rise phenomena were actually to be experienced.

Indirectly cooled stations utilising cooling towers are normally sited inland and would not be subject to this increased risk of flooding. In efficiency terms however, they too would lose cooling performance due to the increased ambient air temperature regime. The physics governing this loss is much more complicated for towers, wherein the bulk of the heat transfer is through evaporation, rather than conduction and convection. Essentially the higher temperature of the cooling air would reduce the "approach", a critical thermal design parameter marking the difference between the wet bulb temperature of that incoming air and the design temperature requirement for the re-cooled water emerging from the tower. Evaporative cooling would be substantially reduced by any reduction in this approach temperature, as would be caused by our $4°C$ rise postulation. Recalling that the full cooling range of a typical evaporative tower rarely exceeds $10°C$, readily illustrates the scale of the likely loss. Again, as with directly cooled stations, the whole, damaging positive feed back loop of reduced efficiency, more compensatory fuel burn and more greenhouse gas production would thus be established.

To describe these cooling problems caused by climatic change in a balanced way, it should noted that revised system designs for the changed climatic scenario can readily achieve the normal cooling efficiency levels at relatively little additional capital cost —a good example of my earlier "designing out" concept.

Siting of tidal power stations and wave devices

If we rashly presume that governments do eventually introduce the concept of making the polluter pay, through carbon taxes or whatever other political method is chosen to reflect the all-in price of each generating system fairly, including environmental damage compensation, the present commercial plight of the clean renewable technologies could be relieved. They could then become serious contenders for power generation, offering the obvious advantage of not adding to the burden of global warming.

However, even given that improbably utopian change from the short termism so beloved by the west's financial institutions, the postulated sea level rises in our scenario today would present special difficulties for two of the leading contenders, tidal power and coastal wave power. Ironically this results from their unusual advantage of offering at least a century of working life rather than the 30 – 40 years of the conventional power station. On present speculation, that century may well encompass sea level increases well beyond today's scenario figures, making the selection of the critical design levels for sluices, turbines, caissons and embankments a series of variables, if their performance is to remain reasonably optimised throughout. Such adjustments throughout operational life would add to the overall costs of such devices, partly negating their advantage of minimal operational and maintenance costs over an extremely long working life.

Hydro-electric stations with increased precipitation

My final example of the possible energy related effects of our climatic change scenario figures is an obvious one of gain. Although the proportion of hydro-electric generation in the UK is small, it does offer added value through being rapidly available on demand and by being so easily controllable. Hence any additional water storage, through our postulated extra 10% of winter precipitation, is obviously a doubly welcome boost to this environmentally friendly resource. Reservoir control measures may be required to optimise the advantage of the extra water, but it is hard to envisage a situation in which the hydro operator would allow this gift to simply spill away.

Unfortunately this extra water would be of minimal value to pumped storage hydro stations, which effectively operate as closed circuit systems, largely unaffected by rainfall since their catchments are usually small.

Shoreline management in response to climatic change

C. A. FLEMING, PhD, FRGS, FICE, Director, Sir William
Halcrow & Partners Ltd

*There are many uncertainties associated with climatic change as it is likely
to impact on the coastline. This requires shoreline management strategies
to be responsive to continued and anticipated change. Given that we have
been experiencing both sea level changes and wave climate in the past it
is necessary to identify with confidence cause and effect of coastal
changes. The means by which this can be achieved is through systematic
and consistent monitoring of the shoreline. This will then provide the
information that is necessary to develop shoreline management response
to address anticipated events as well as to assess properly the rate at which
changes are really taking place.*

Introduction

This paper concerns the planning and management of the shoreline, its
beaches, coastal defences and immediate foreshore and hinterland insofar
as they interact with those measures that need to be taken to maintain the
shoreline in a manner that serves all of the needs of the community. It is
therefore, by definition, a sub-set of coastal zone management in its widest
sense that might involve a significantly broader range of issues. The
shoreline is an extremely sensitive area whose behaviour in response to
changes in both distant and local conditions is potentially quite dramatic.
These conditions principally include mean sea level rise, severity and
direction of storms and magnitude of storm surge levels. Thus, changes
on a coastline result from the interaction of sea level fluctuations, wind
induced phenomena, uplift/subsidence of the land mass and variation in
the sediment supply. These all tend to be episodic in nature with differing
rates and timescales. Consequently, changes take place over periods which
vary from days to several decades, with relevant geological changes
occurring over even longer periods.

Over recent years there have been some fundamental changes in approach to the provision of flood protection and measures to limit erosion.

Engineering for climatic change. Thomas Telford, London, 1993

In practical terms this has resulted in new types of structures and in particular a 'soft' approach, based on techniques of beach nourishment and subsequent management, as an alternative to the more traditional 'hard' approach, using sea walls and groyne systems. Implicit in the use of such alternatives is a greater understanding of coastal processes, which has come about as a consequence of international research efforts over the past few decades. It has also been suggested (Ref. 1) that soft defences which have shorter design lives are more easily adaptable to changing boundary conditions.

As such techniques are based on understanding the dominant coastal processes of an area, it has become important to study the whole coastal regime rather than just the frontage of interest. It has become clear from these more regional studies that there are great benefits to be gained from adopting a strategic approach to the provision of coastal works. At the same time, there are increasing pressures from the diverse and often conflicting range of coastal interests, such as conservation, tourism and industry. This therefore requires a more complete assessment of the shoreline management options and has provided the incentive to develop suitable management techniques.

The major difficulty that arises at present is that, whilst the rate of sea level rise has been estimated to a confidence limit that can at least, pro tem, be accommodated until more detailed information becomes available, the more significant aspects of potential changes in storm intensity and direction are not well established and are open to considerable uncertainty. Knowledge of trends on the shoreline can only be improved through systematic monitoring of both the cause and effect of changing conditions.

Impact of climate change
Coastal forcing

The phenomena that are the primary cause of natural coastal change are sea level, land uplift/subsidence, wave climate, storm surge and sediment supply. The authoritative source of climate change parameters is provided by the Intergovernmental Panel on Climate Change (IPCC, Ref. 2). It is also generally considered that the 'business-as-usual' scenario should be used as the basis for predicting future trends. These are themselves based on the outcome suggested by state-of-the-art global circulation models.

On this basis the best estimate of global sea level rise is given as an increase of 200 mm by the year 2030 increasing to 300 mm by the year 2050. These values generally apply to the UK but must be adjusted locally to allow for land subsidence or uplift as well as tide conditions which not only vary from place to place, but also may change in response to sea level rise itself. Where water depths increase this could allow the tidal wave to propagate faster, thus altering currents and levels. This may be counteracted by rapid morphological changes which will naturally tend to compensate for such effects. it has also been noted that, even disregarding the question of increased storminess discussed later in the text, an increase in sea level will tend to extend the area of the mild offshore slope resulting in an increase in surge elevations. However, where the shelf is of uniform depth, there could be a decrease in wind induced surge (Ref. 3).

Changes in wind strength and pattern can potentially effect the wave regime in a number of ways. A general rise in wind speeds would increase wave heights for ambient conditions. At the same time dominant wave directions might also change. It has also suggested that climate changes associated with the greenhouse effect may alter the storminess of particular sea areas, possibly as the result of deeper depressions or some change in the general tracking of depressions. Whilst such changes have not been directly associated with the greenhouse effect it has been suggested that wave heights in the North Sea and North Atlantic are increasing (Ref. 4).

One of the many studies carried out as part of the Anglian Sea Defence Management Study (ASDMS) examined what predictions could be made with respect to increased storminess in the North Sea (Ref. 5). Firstly the study investigated the storm climate during the last century by examining the storm climate in terms of the daily pressure field for the North Sea to create what is known as a gale index. These were ranked into classes of gale, severe gale and very severe gale over the period 1881–1989. These were also related to more direct parameters of wave height and surge level on a limited number of data sets. The mean seasonal circulation based on observations and the predictions of four global circulation models (GCMs) were compared for different CO_2 (the dominant greenhouse gas) scenarios that might be relevant in the future. The results indicated no long-term periodicity in North Sea gales, but suggested that, following a 60 year period of very gradual decline in the number of gales per year, the past 20 years has seen a steady increase in the number and intensity of gales. It also appeared that the intensity of gales occurring in a particular year would most probably be followed by a relatively calm year.

The second part of the analysis examined the relationship between the gale index, maximum significant wave height and storm surges. This indicated that increased storminess is associated with increased surge levels, but requires further refinement with more data to account for the residual variance in surge height. Computer models were used to assess the nature of storminess in the next century over the North Sea on the basis of one and two times the present level of the predominant greenhouse gas. This was used to construct a gale index for the 21st century that could be directly compared to the gale index derived from historical data. The results indicated a slight increase in storminess during the next century, but that this would be within the bands of natural variability. The conclusion was that, provided that background changes in sea level are taken into account, the forcing conditions for waves and surges over the next one hundred years will not exceed past experience. At the same time it is acknowledged that the predictive capability of GCMs are at present limited in their ability to represent meteorological phenomena accurately at a local scale.

Coastal response

The impact of climatic change on beaches has been addressed in another paper (Ref. 6). In general terms the concept of an equilibrium beach profile, in response to prevailing wave conditions has been established for some time. As sea level rises, it has been demonstrated for non-cohesive beaches that the beach profile is translated shorewards with recession of the upper beach and deposition on the lower foreshore. For cohesive shores, which may comprise of a sand veneer beach overlying a clay sub-base it has been suggested for shorelines in the Great Lakes that they also exhibit an equilibrium profile and that this can be related to uniform energy dissipation across the surf zone. However, data collected and analysed during the ASDMS showed that cohesive beaches have both retreated and steepened. The essential difference between the two areas is that the east coast is subject to a significant tidal range coupled with long term sea level rise whilst the Great Lakes have no tides but have experienced significant water level changes.

Thus, the impact of sea level rise on beach profile response can be readily anticipated, but only in two-dimensional terms. The potential impact of wave climate change, whether in terms of ambient or extreme conditions, is enormous. The alongshore drift on a coastline is a non-linear function of the breaking wave height and its direction with respect to the beach normal. Simple calculations show that small changes in either the wave height or direction result in significant changes in drift rates. It follows that if these are accumulated over a period of say a year there is the potential for major changes in net alongshore drift volumes and consequently the distribution of beach material along the shoreline due either to natural geophysical features or to man-made structures.

The normal pattern of beach response following a storm is that of recovery of material that has gone offshore. That recovery process is over a time span rather greater than that of the event itself. It follows that an increase in frequency of storms could lead to more instances when a beach fails to recover before being subject to further damage and so on. Increases in both the level and frequency of storm surges could also have significant impacts by allowing larger waves closer into the shoreline with less energy dissipation on the approach. Where there are cliffs or hard defences at the top of the beach increased reflection could cause further damage.

Marshes and saltings can respond readily to sea level rise as long as there is sufficient sediment supply and the changes do not occur too rapidly. Barrier beaches can be expected to roll back as sea level rises. This may result in a reduction in the area of marsh unless it is also able to migrate landwards. In some areas this would require the existing sea defences to be abandoned to provIde that space.

Coastal structures will also be subject to changing conditions of water level, wave height and direction. If the back shore is lost or water depths in front of a sea wall or revetment increase, the effect on the wall's performance in terms of run-up and overtopping, as well as the stability of the armour, could be significant. Other coastal structures such as groynes, headlands and offshore breakwaters would also require similar considerations.

Shoreline management

In order to be able to contemplate a shoreline management approach it is necessary to obtain a proper understanding of the natural processes involved. For a given regime or region, studies will necessarily consider historical changes, the effects of the marine climate, the patterns of sediment movement and probably need to assess water quality and eco-logical sensitivity. This then allows the management issues, such as conflicts of interest, level of protection and type of protection to be considered in the light of a proper understanding.

Nevertheless, whether considering issues or processes, the fundamental requirement is information. The balance between the two will usually only alter the range and extent of data collected. Thus, the principal input is information, whether existing, generated by routine monitoring, or the result of forecasts.

The information on processes and usage provide a fundamental input to the development of a management strategy. Such a strategy must obviously be formulated within the constraints of the legal and institu-tional frameworks which prevail. It will therefore comprise objectives based on the legal obligations and aims of the authority. These will be

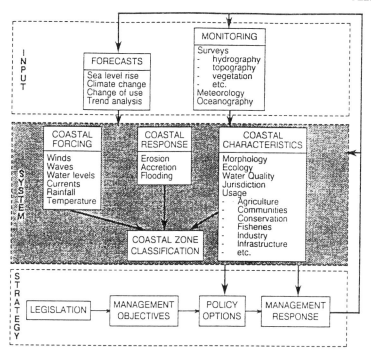

Fig. 1. Responsive management framework

taken forward by policy guidelines which identify both the imposed constraints and strategic decisions which endeavour to make best use of available resources. Whilst these guidelines should assist future planning, the management response options define how the various options can be implemented. Taken together the objectives, guidelines and response comprise the management strategy.

As part of the ASDMS a Management Framework was developed and has been fully described elsewhere (Refs 5, 7, 8). It is evident that in order to maintain an up to date picture of the factors which effect coastal zone classification and ultimately management response it is necessary to maintain a dynamic system as shown in Fig. 1. It can be appreciated that this approach is even more essential when such uncertainties as sea level rise, climate change and consequent shoreline response exist.

This 'responsive management framework' requires the coastal classification to be updated based on the results of routine monitoring and/or the output of forecasts. As this changes so does the management response strategy which in turn feeds back by influencing the coastal characteristics through the implementation of coastal works. A programme of routine monitoring will provide data on winds, water levels, beach profiles and coastal habitats, which can be processed and reduced to a form that can be added to the management system. The shoreline manager can then use the system to re-evaluate the coastal classification by re-working the interpretive analysis.

In a similar way, studies of change can be initiated, such as a change of use, or changes due to forecast sea level rise or climate change. These can be carried out as independent exercises and the results compiled in terms of a new definition of coastal characteristics, forcing and response. This new set of parameters can then be introduced and the coastal classification and hence management response can be re-examined. Hence the framework provides a means of responding to changes, once these changes have been defined.

Shoreline monitoring

It can readily be appreciated that an essential key to determining the impact of climate change on the shoreline, which is already taking place, is to systematically measure and record the changes that are occurring. It is most unfortunate that in the majority of cases historic data is sadly lacking and there is little to base forecasts upon.

The effectiveness of any management process is determined by the accuracy of the information upon which decisions are to be based. The information gathered in the ASDMS has provided much new understanding of coastal processes. This has already led to a more regional approach to solving the flood defence problems of particular lengths of coast. These very developments and the changing nature of coastal processes themselves mean that the database will soon lose its value unless the information is kept current. This means that data on the most essential variables must be updated on a routine basis as part of a structured monitoring programme.

A comprehensive monitoring strategy has been developed which has resulted from careful studies which identified both the nature of data required and the frequency at which it needs collecting in order to provide the most effective information required by the shoreline manager. The programme focuses on the forcing components and response component as described in Tables 1 and 2 (Ref. 9). The scope described provides the basis for collecting relevant data without placing an undue burden on financial resources of an authority.

Clearly simply collecting data and archiving it is of little value as one

Table 1. Data requirements for forcing component

Description	Format	Frequency	Analysis	Use
Wind	Velocity and direction	Hourly	• Inshore wave climate • Offshore storm climate • Extremes	To assess shoreline exposure
Water Levels	Height ref to datum	At least hourly	• Tides • Extremes • Joint wave and water level analysis	Required for inshore wave computations and allows significance of storm surge to be identified
Tidal Prism	Water levels and currents	Every 5-10 years	Estuary power curve	To define estuary dynamics and in the longer term to provide indication of changes in the hydraulic regime

of the main objectives of a monitoring programme is to get some fairly immediate feedback on shoreline behaviour and structural integrity of defences. With this in mind a map based data storage and analysis system has been developed to provide a wide range of functions, as shown in Fig. 2. The core function allows beach inspections, structure inspections, beach

Table 2. Data requirements for response component

Description	Format	Frequency	Analysis	Use
Aerial Survey	Photographs	Annual	Foreshore and backshore levels	To identify key features, quantify change in both a plan and cross-shore sense and to monitor both physical and ecological changes
Hydrographic Survey	Depths along profile lines	Every 5-10 years	Changes with time	Establish limit of sediment exchange on shore face and provide detailed, up-to-date bathymetry for process studies
Land Survey	Levels along profile lines	Biannual	Changes with time	Allows changes in the beach to be studied and related to forcing data (Table 2)
Inspections	Text records	Biannual	Changes with time	Observation of beach features and coastal structures can be related to processes and key events, to help establish patterns of shoreline behaviour.

89

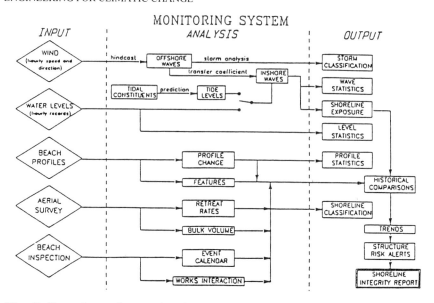

Fig. 2. Functions of a monitoring system

profiles and miscellaneous events to be recorded in a consistent manner and against a time reference. Supporting functions allow time series of winds, waves, tide surges and beach profiles to be compared on a synchronised timescale. At the same time statistical analyses allow links between measured processes that affect the coast and date from surveys of the shoreline's response.

It can be appreciated that, as the timescale which the data cover expands, so does the ability to detect cyclical or monotonic trends. For example, comparison of a particular beach profile with the long term maxima and minima mean and statistical trend immediately provides an indication of the relative state of the beach at any one time. This then needs to be coupled with the recent and long term history of forcing functions. For example, a storm typicality energy matrix compares the actual number of storms in each energy category which have occurred during a specified data range (e.g. since the previous beach profile) with the 'typical' number of storms over the same seasonal time period from the complete time series available for a particular location. Similarly a storm typicality direction matrix provides a measure of the actual and typical storms that have occurred within pre-defined direction sectors. The difference between the two provides an assessment of whether the direction of storms in the period of interest differs from expected values, based on the long term record. Combined with the storm energy matrix it is possible to assess variations in both storm severity and direction. When linked to the measure of coastal response, i.e. the beach profile, a judgement can be made as to whether a severely degraded beach is the result of one or more unusual events or whether some other cause, such as reduction in sediment supply, needs to be sought. Similarly where coastal defence structures are involved records of condition, overtopping, toe scour etc. can be linked with recent and longer term trends.

Conclusions

Much is talked about regarding the uncertainty of climatic change and the threat that it implies. However, it should not be forgotten that we have already been experiencing sea level rise for some considerable time and we have also experienced wave climate variability in terms of storm frequency over periods of several decades. The difficulty is that we are not precisely sure what the impact has been in the past, and are even less sure as to what might be experienced in the future. It is for this reason that it has been suggested that response to such changes should be rational and progressive and we require sufficient understanding of the coastal environment to allow estimates of flood probabilities and their changes commensurate with the likely lifetime of the defences (Ref. 1). Designs must be developed that can cope with changing conditions. However, more fundamentally it is necessary to develop long term strategies that address the more difficult decisions to possibly abandon defences in particular areas as part of a managed retreat programme.

Management tools have been developed that allow shoreline management strategies to be evolved, leading to the appropriate management response, but these decisions can only be made on the basis of information that is currently available. This is generally inadequate to address the problems described. There is therefore a strong case for initiating comprehensive monitoring programmes in order to accumulate the type and quality of data that is necessary to go forward confidently with shoreline management strategies. The significant benefits that can be derived are that

- it forms an essential link within the overall shoreline management framework, which ensures that due account is taken of the dynamic nature of the coast in formulating the management strategy;

- the move towards soft engineering in the form of beach recharge, etc., requires a greater commitment to regular monitoring;

- monitoring on a regional basis will enable patterns of change to be identified and is therefore likely to provide an early indication of the direct impacts of climatic change;

- the long term collection of data will provide a sound basis for defining design parameters, as well as the variability of individual parameters. The latter will be needed for any form of risk analysis, required to establish the standard of defence being provided.

Acknowledgements

The author would like to thank his colleague Ian Townend who provided some of the material contained in this paper.

References

1. Department of the Environment. *The Potential Effects of Climatic Change in the United Kingdom.* HMS0. 1991.
2. HOUGHTON J. T., JENKINS G. J. and EPHRAUMS J. J. *Climatic Change. The IPCC Scientific Assessment.* Cambridge University Press. 1990.
3. NATIONAL RESEARCH COUNCIL, CANADA. *Responding to Changes in Sea Level: Engineering Implications.* National Academy Press, Washington. 1987.
4. TOWNEND I. H. *Effects of Sea Level Rise on the Coastal Zone. The Greenhouse Effect and Rising Sea Levels in the UK.* MAFF Conference 1989. ed. J. C. Doornkamp, M1 Press Limited.
5. NATIONAL RIVERS AUTHORITY. *The Future of Shoreline Management Conference Papers.* Sir William Halcrow & Partners, NRA, Anglian Region. 1991.
6. HUNTINGTON S. W. and POWELL K. A. *Management of beaches in an environment of changing sea levels and wave climate.* This volume.
7. FLEMING C.A. The Anglian Sea Defence Management Study.*Coastal Management.* 1989. Thomas Telford, London.
8. TOWNEND I. H. 1990. Frameworks for Shoreline Management. *PIANC Bulletin* No. 71, pp. 72 – 80.
9. BARBER P. C. and TOWNEND I. H. *Monitoring Guidelines.* 1991. Report for the National Rivers Authority, Anglian Region.

Appropriate solutions for maritime engineering

R.S. THOMAS, Director of Coastal and River Engineering, Posford Duvivier and M.W. CHILD, Public Relations Manager, National Rivers Authority, Anglian Region

Introduction
The stated aims of the conference which led to this volume were to "clarify the implications" (of climate change) "for all civil engineers and to discuss what actions should be taken now". The implications of climate change for coastal defence and port and harbour authorities, in fact for all those related with coastal civil engineering, are potentially

- the need for increased investment or the acceptance of declining standards

- the need for a strategic plan to deal, not only with the potential increases in sea level and greater severity of wave action, but also with the uncertainty surrounding the predictions of change.

Papers in this and other volumes and symposia have dealt with the possible consequences of climate change, which are likely to alter the wave climate and tidal streams. Deeper water resulting from higher sea levels will allow larger waves to reach the shore and potentially, new or revised flow paths for tidal streams to develop. The higher waves inshore may well be compounded if increased storminess produces a more severe wave climate offshore. In order to develop responsive strategies for maritime engineering works, the engineer should have as thorough an appreciation as possible of these aspects and this in turn requires a good understanding of coastal processes — processes such as wave induced and tidally induced sediment transport. In such a way the coastal engineer will have a basis on which he can assess the impact of climate change, for example with respect to

- changes in sedimentation in rivers and estuaries — quite small changes in mean sea level can bring substantial increases in sediment deposition in rivers

- changes in erosion rates on the coast, whether by virtue simply of natural retreat up the coastal slope, or by the dying back of coastal vegetation

- direct effects of increased water level

- direct effects of a more severe inshore wave climate.

These must all be viewed in the context that, in all honesty, we do not have much idea at all as to the extent of the acceleration of sea level rise or, indeed, as to the existence at all of the increased severity of wave action. The engineer needs to understand the sensitivity of his designs to possible changes in ambient and extreme conditions. This should not, however, simply be used as a reason for overdesigning.

Rather, it should he seen as a situation where designs should be produced which are not oversensitive to changes in prevailing conditions and which can be upgraded or adapted without major reconstruction or obsolescence.

For the promoter, or client, the key to the future is to know how and when to commit funds against the uncertainties of climate change. For the designer the key is flexibility.

The need for a strategy

The maritime engineering environment is aggressive and subject to complex changes that are not fully understood. The prospect of climatic change and sea level rise further compounds this already complex and unknown environment. Additionally, most maritime engineering is carried out at the shoreline, the sensitive interface between the land and sea, an area of high landscape value, of coastal heritage and amenity, and conservation sites for flora and fauna. These aspects, and the need for an integrated approach to coastal development, maritime engineering, coastal usage and conservation, are well documented in various reports on coastal zone management (Ref. 1) and were recently debated in the House of Commons Environment Select Committee Report (Ref. 2) on coastal zone protection and planning. Climate change has not created the need for a coastal zone management approach it has merely emphasised the importance of strategic and integrated planning in the complex coastal environment.

Maritime engineers have a responsibility not just for the mechanics of a particular design proposal but also for the wider view and strategic implications of their proposals. This wider view increasingly will rely on understanding of processes which in turn will rely on the development of numerical models and monitoring of the coast and of projects as they are completed. The challenge for the future is to accommodate the uncertainty of climate change in strategic planning and project design and development, to ensure cost effective and environmentally acceptable solutions are implemented.

Flood defence

A good example of the need for strategic planning is flood defence. The NRA maintains more than 3000 km of sea and tidal defences in England and Wales. In 1990/91 £46M supported by MAFF funding was spent on new capital projects and this is forecast to exceed £60M in 1993/94. The Anglian Region of the NRA is responsible for the largest area of flood risk (Fig. 1). During the 1953 floods over 300 people were drowned and 20 000 homes destroyed or damaged in the region. It has been estimated that the current £30M annual investment in the Anglian Region will need to be increased four-fold for a period of at least twenty years just to increase the height of defences to cope with the sea level forecast for the year 2050. Changes in storminess and erosion rates caused by climatic change would further increases this investment.

The NRA has the responsibility to provide flood defences where economically justified and within available benefits, for people, property and land. It is a responsibility that rightly attracts a high media profile, particularly after a flooding event and during this period of uncertainty about climate change and sea level rise. Media attention and public concern, however intense and well meaning, does not negate the need for a strategic plan to properly assess the risks, uncertainties, options and costs of protection against sea level rise.

In developing the strategy, some of the questions to be answered in relation to risk and uncertainty are as follows:

1. Timing — Where should public investment be made? Should we risk waiting until evidence is conclusive? Should we invest now and risk abortive expenditure?

2. Standards of protection — Should we improve all defences to maintain the current level of protection? Should we accept a reduction in the standard of protection in some areas? Should we abandon some areas?

95

Fig. 1. Anglian region flood risk area

3. Methods of protection — How should we protect people and property? Which are the best methods? Which methods are less sensitive to future sea level rise? Which are the best methods to use taking into account local geology, processes and wave climate? Which methods are least damaging to the natural sediment budget?

Clearly there is not one single solution to all these imponderables. The strategy must be broadly based and flexible enough to accommodate future change and improvement in knowledge and understanding of climate change and coastal processes. The essence of the strategy includes

- incorporation of allowances for predicted sea level rise in new designs (Ref. 3)

- ensuring that at the project appraisal stage design options are tested for sensitivity with respect to sea level rise, and that they can be upgraded or adapted without major construction

- looking at the coastline as a whole and improving knowledge and understanding of coastal process (Ref. 4)

- monitoring the coastline and developing a geographical information system (GIS) database (Ref. 5)

- investing in research and development to determine the most suitable designs and developing soft engineering techniques (Refs 5 –7)

- collecting and analysing data on defences, including condition assessment and standard of service (Ref. 8)

- developing shoreline management plans (Ref. 9)

This strategic approach is well documented in the references mentioned above and is largely based on the work carried out by Halcrow on the Anglian Region (NRA) Sea Defence Management Study (Ref.10) which commenced in 1989 and was completed in 1991. A early task of this study was to understand coastal processes and develop a GIS database for the management of data. The major future tasks are
- monitoring the coast and collection of quality data
- developing a database of standards of service based on risk assessment
- development of shoreline management plans.

Monitoring
    A key development of the Sea Defence Management Study and the cornerstone of the strategy is the collection of quality and relevant data. This will help in the early identification of sea level rise from climatic change and evaluation of future changes to tidal streams and erosion rates. Without this knowledge today the maritime engineer will find it difficult, if not impossible, to make rational decisions tomorrow about the impacts of climatic change.

97

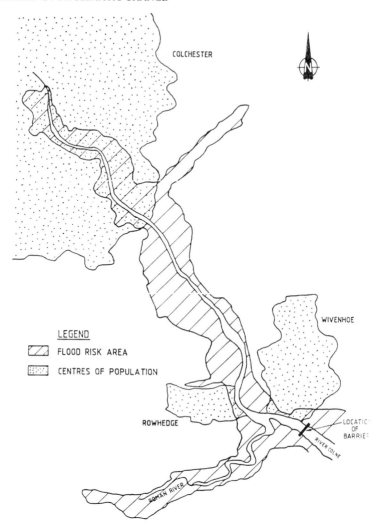

Fig. 2. Area protected by Colne barrier

Flexibility in design

We have already introduced the aspect of uncertainty and the consequent need for flexibility in engineering. These points are perhaps best illustrated by examples.

A surge barrier

Colne Barrier is being built for NRA Anglian Region to provide flood defence against extreme high tides for much of Colchester, Wivenhoe and

Rowhedge - many tens of millions of pounds' worth of property on extensive low lying land (see Fig. 2). Essentially two main options for flood defence were available for consideration (Ref. 11). These were

- a surge barrier

- bank raising (i.e. conventional food defences along the river's edge).

In evaluating the sensitivity of designs, in the uncertainty associated with climate change, a surge barrier offered flexibility in many areas which strongly militated in its favour. These were

- Port operation — Colchester is a working commercial port. Flood defence through bank raising would have involved quayline flood defences. If they became more than a metre high, operation of the quays would be difficult. The quaysides would need to be raised (necessitating new quay walls) in order to allow continued operation. Costs could have been an order of magnitude greater if sea level rise had been much greater than anticipated.

- Visual impact — The villages of Rowhedge and Wivenhoe border the River Colne and at present have no flood defences. Provision of defences to a 100 year standard now would need flood walls of the order of a metre high. This would he a substantial physical intrusion into the presently scenic waterfronts, but could be landscaped. Sea level rise of, say, 1m would necessitate walls 2 m high which would destroy the character of the frontage and the views of waterside properties.

The above points demonstrated the inflexibility of bank raising on the River Colne. On the other hand, a barrier offered considerable flexibility, for example

- Raising of the defences — in the event of an acceleration in sea level rise was concentrated in a small area, i.e. at the barrier. Problems such as those outlined above were avoided. Actual raising of the barrier by of the order of 0.3 m can be effected without serious lowering of factors of safety, and at minimal cost. The ability of the Barrier to accommodate such changes was designed into the original structure at little extra cost.

- Navigation — the barrier will only close on those tides likely otherwise to cause flooding upstream. It is envisaged that it will close some four times per year. At this frequency it does not pose a serious nuisance to commercial shippers and other river users. In the event of a more rapid rise in sea level, NRA will need to review the situation. They may choose either to close the barrier more frequently, or to build limited flood defences upstream so that the number of closures per annum remains at four. The barrier provides them with the flexibility to choose depending on river usage at that time.

Rubble breakwaters

The construction of 'rubble' breakwaters is the most common form of breakwater construction in Europe. Typically the construction consists of a rubble core armoured by between two and four layers of rock progressively larger in size, culminating in a layer of armour large enough to resist storm wave action. These outer layers would normally be laid at a slope of 1:1½ to 1:3. As an alternative to large rock, armour units (e.g. dollosse or accropodes) can be used. By virtue of improved interlock, armour units laid (perhaps in a single layer in the case of accropodes) at a steep slope can offer greater resistance to wave attack than much larger rock at a flatter slope.

Rock armour functions by being sufficiently massive to resist the impact of waves without moving. There is little or no interlock, although the angularity of the rock has an impact on stability (i.e. angular rock is more stable than spherical rocks by virtue of increased friction).

Concrete armour units were developed to allow the construction of rubble breakwaters in areas where there was insufficient rock of adequate size to withstand the local wave climate, i.e. in areas where the wave climate is very severe (especially when constructing in deep water) or where large rock cannot be obtained. The stability of armour can be characterised by the stability factor K_D. K_D for armour rock is in the region of 3, whereas for some armour units it is in excess of 15 (Ref. 12). This means that armour units can sometimes be used as an alternative to rock which is five times heavier. Many armour units can, in addition to this, be laid at a steeper slope — say 1:1/3. Armour units therefore offer the potential for using less material, or a smaller size, which can lead to substantial costs savings.

When we compare the use of armour rock with the use of armour units in the context of uncertainty of climate change, the potential advantages of armour units may not always apply. Randomly placed armour units, such as dolosse, achieve their great stability through interlock between units. The units tend to be 'leggy'.

Their shape means that the stresses induced by self weight can be

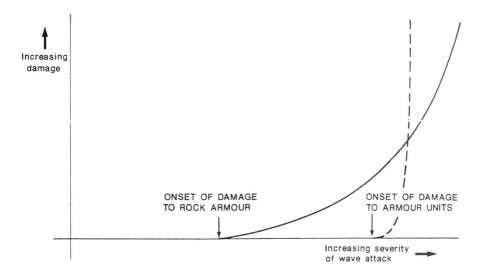

Fig. 3. Variation of damage with wave climate (indicative only)

significant with respect to allowable tensile stresses and this situation is exacerbated by interlock, which further increases tensile stresses, leaving a very small margin before breakage occurs. Furthermore, when design wave conditions are exceeded and units begin to move the interlock is broken and the units begin to behave more like rock, i.e. they may re-stabilise at a flatter slope governed by mass of the units and inter-unit friction.

The consequence of potential breakage and movement of units is that breakwaters armoured with armour units can exhibit a more brittle failure than rock armoured structures, as indicated in Fig. 3. The onset of damage to rock armour may occur with milder wave conditions than for armour units, but once the armour units are damaged total collapse may rapidly ensue. In circumstances where we are not sure how wave climate will change in the future, armour units can be more sensitive to changes than rock. That is not to say that we should not use armour units. Uncertainties abound in coastal engineering. Some apply to rock and not to armour units. Uncertainties abound in coastal engineering. Some apply to rock and not to armour units. This aspect is just one of the uncertainties. We should, however, check what will happen if presently envisaged extreme conditions are exceeded. This, it should be remembered, should be an important element of breakwater design in any case, so that designers are aware of the significance of errors in extreme value estimates.

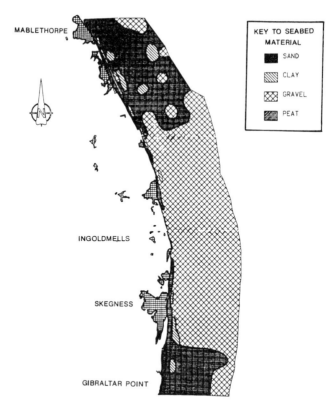

Fig. 4. Mablethorpe Skegness defences: coast covered by study

Beach nourishment

The sea defences between Mablethorpe and Skegness extend for some 24 km of the Lincolnshire coast (Fig. 4) and protect large areas of low lying land and property (for example, more than 15 000 homes). Historically, and particularly since 1953, the defences have been sea walls, whether revetments, concrete stepwork or rock slopes. NRA (Anglian Region) are nearing the end of a major reconstruction programme to provide secure flood defence for Lincolnshire and in 1991 took the opportunity to carry out a detailed and extensive review of their future policy towards defence of the area. The outcome of this was a decision to implement a major programme of beach nourishment, commencing in 1993 (Ref. 13).

The beaches fronting the seawalls have been falling in level for many years (i.e. at least since 1953 when records began and probably for much longer). The beaches consists of a thin veneer of sand overlying clay, and

Table 1. Present value of costs (£millions) of seawall and beach nourishment options

	Seawall	Beach
Sea level rise of 5mm pa.	104.4	80.3
Sea level rise of 8mm pa.	108.6	80.7
Increase in costs due to greater sea level rise	4.2	0.4

it is the wave induced erosion of this clay which has caused the beaches to fall. The provision of a much larger beach (amounting to some 400 m^3 extra sand per metre of defence), with subsequent re-nourishment, will to a large extent prevent further erosion of the clay and will protect the hard defences from all but the most severe storms, thus minimising the cost of future maintenance of the seawalls. It will provide for secure defences in the future as well as preserving a valuable amenity. Most importantly, in the terms of this volume, it will give a system of defence that in this location will be relatively adaptable to future climate change.

As mentioned in the introduction to this paper, the consequences of climate change could well be larger waves inshore, both in terms of a generally more severe wave climate, and in terms of deeper water which will allow larger waves to reach the shore. Reliance on seawalls may well necessitate seawall strengthening, raising, or reconstruction earlier than envisaged at design. In the strategy study for the Lincolnshire defences, the sensitivity of the two options (seawalls and beach nourishment) to differing rates of sea level rise was assessed.

In the study a future sea level rise of 5 mm per annum was assumed, being considered the best estimate at that time. Costs were derived for both options, including future work, for the 50 year life of the scheme. Costs were also derived for achieving the same standards of defence but assuming a sea level including future work, for the 50 year life of the scheme. Costs were also derived for achieving the same standards of defence but assuming a sea level rise of 8 mm per annum. When discounted to 1991 (the date of the study) the results in Table 1 were produced.

The assumptions made in this sensitivity test were somewhat crude and did not allow for any increase in storminess. None the less it is clear that, in financial terms, the use of beach nourishment has considerable advantages.

Conclusions

In this paper we have demonstrated that the key to engineering in the uncertainty of climatic change lies in understanding the processes at work and then designing so that structures and other works of engineering are flexible. Examples of how this can be achieved have been given. On this basis certain conclusions can be drawn.

(i) There is a clear need for appropriate, well planned and continuing monitoring which assists in our understanding and helps in the early identification of the consequences of climate change.

(ii) Solutions can be developed which are less sensitive to climate change than others. Sensitivity tests are an essential part of designing in the uncertainty relating to climatic change.

Designing in this context of uncertainty, however, should be nothing new to the coastal engineer. Perhaps the public perception of climatic change allows him to be a little less reticent about admitting it. Design and construction in the sea is riddled with randomness, statistical sampling, variation in materials and complexity, and even includes the British weather. In this context, the need for understanding, monitoring and sensitivity testing has always been fundamental to good coastal design (Ref. 14). Climate change merely emphasises the need for it.

References

1. GUBBAY S. *A Future for the Coast?* Proposals for a UK Coastal Zone Management Plan. Marine Conservation Society, 1990.
2. HOUSE OF COMMONS ENVIRONMENTAL COMMITTEE. *Enquiry on Coastal zone protection and planning,* 1992.
3. SWINNERTON C.J. *Engineering in the uncertainty of climatic change: the view of the NRA.* This volume.
4. FLEMING C.A. The Anglian Sea Defence Management Study. *Coastal Management.* Thomas Telford, London, 1990.
5. CHILD M.W. *The Anglian Management Study.* Thomas Telford, London, 1992.
6. MINISTRY OF AGRICULTURE FISHERIES AND FOOD. *Flood and Coastal Defence Research and Development,* MAFF, 1992.

7. NATIONAL RIVERS AUTHORITY. *Review of coastal and estuarial R&D related to Flood Defence*. Project report 308/2/HO, 1991.
8. NATIONAL RIVERS AUTHORITY. *Sea Defence Survey.* Leaflet,1992.
9. TOWNEND I. H. Shoreline Management: A Question of Definition. *Coastal Management.* Thomas Telford, London, 1992.
10. NATIONAL RIVERS AUTHORITY and SIR WILLIAM HALCROW AND PARTNERS. *Shoreline Management; the Anglian Perspective*. Conference papers of the Future of Shoreline Management Symposium, October 1991.
11. POSFORD DUVIVIER. *Upper Colne Estuary Tidal Defences Feasibility Study*. Final Report, on behalf of Anglian Water NRA Unit, 1988.
12. US ARMY CORPS OF ENGINEERS. Shore Protection Manual, 1984.
13. POSFORD DUVIVIER *Mablethorpe to Skegness Sea Defences — Strategy Study*. Study for NRA Anglian Region, November 1981.
14. THOMAS R.S., HALL B. and the CIRIA. *Seawall Design*, Butterworth-Heinemann, London, 1992.

Management of beaches in an environment of changing sea levels and wave climate

Dr S. W. HUNTINGTON and Dr K.A. POWELL, HR Wallingford Ltd

Much of the United Kingdom coastline is protected, at least in part, from flooding or erosion by the presence of beaches. If this protection is to be maintained it is necessary to ensure a full understanding of the response of a beach to a scenario of future climate change, and to incorporate measures to deal with that response in any future beach management strategy. This paper describes the range of beach types found around the coastline and their potential responses to changing sea levels and wave climate. The implication of these responses is examined and it is concluded that whilst there is unlikely to be a need for any radical change in the way in which we approach beach management, there will be a need for a greater monitoring effort and improved predictions of long-term beach behaviour.

1. Introduction

The coastline of the UK, in common with coastlines around the world, has experienced and continues to experience the results of climatic change. The most widely publicised consequence of global warming, namely sea level rise due to increased ice melt and thermal expansion of the upper levels of the ocean, is nothing new. However, many other effects are potentially important, although it is only recently that sufficient information has become available to appreciate their consequences.

Any warming of the Earth and its atmosphere will inevitably bring about changes in the circulation of air around the globe. Such changes could provoke alterations in mean atmospheric pressure and, as a consequence, accentuate sea level variability. Differences between mean sea level at a coast in winter and summer are already considerable in some areas, particularly where wind directions are reasonably constant over long periods of time (months).

The intensity and speed of movement of low pressure areas is of particular concern in the shallow seas of the north-west European continental shelf, because of the generation of tidal surges. A modest increase in the height or frequency of such surges, in the southern North Sea or the Irish Sea in particular, may be much more threatening to beach-based sea defences than a modest long-term rise in mean sea level.

Rapidly moving deep depressions are inevitably accompanied by strong winds, and changes in the strength, frequency and direction of winds are another probable consequence of climate change. The direct effect of wind on coastal defences is only of secondary importance, perhaps increasing spray overtopping or modifying aeolian transport rates and patterns on beaches and over sand dunes. Of greater importance is the effect of changing wind climate on the wave conditions incident on a coastline. These effects may be manifest in three ways:

(i) An increase in the general wind speeds could result in an increase in the general level of wave activity. Since it is usually the more frequent, modest wave conditions which are of greater importance in the transport of sediment along the coast, an increase in general wind speeds could result in a significant increase in sediment transport rates. For longshore transport this effect might prove to be of greater importance than a modest increase in wave heights during a large storm.

(ii) An increase in storm frequency, particularly over southern Britain, will reduce the interval between storms and therefore the time available for a storm-damaged beach to effect a full natural recovery. Since a natural beach can take weeks or even months to recover from a severe storm this could lead to a serious weakening of beach-based coastal defences.

(iii) As wind patterns change it is likely that the dominant wave directions will also change. Where beaches have evolved a planshape in equilibrium with the prevailing wave climate (usually pocket beaches) a slight shift in wave direction will produce a corresponding re-alignment of the beach, which may leave previously protected lengths of coastline exposed to direct wave action. Along most of the UK coastline, however, there is a nett longshore transport of beach material often controlled, or at least heavily influenced, by the construction of coast protection works or harbour arms. On these frontages a slight change in wave direction will modify drift rates and in certain circumstances may result in drift reversal. These changes could leave existing beach control structures isolated and largely ineffective, and will certainly alter existing patterns of beach erosion and accretion.

107

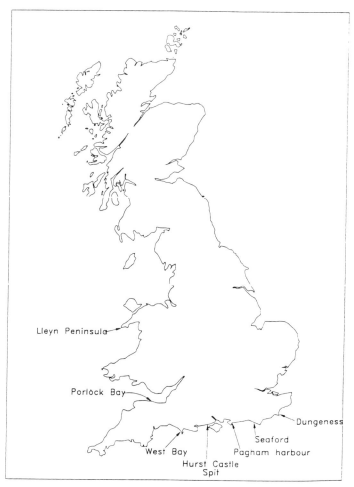

Fig. 1. Location of sites referred to in text

2. Beach types and response to climate change

For the purposes of this paper the wide range of beach types found around the UK coast have been sub-divided into four general categories:

(i) Nett-drift beaches — the most common beach type. Composed of sand or shingle sized sediments subject to a nett longshore drift and backed by high ground or substantial coastal structures.

(ii) Pocket beaches — enclosed beaches, such as those on the Lleyn Peninsula (Fig. 1), which are aligned to the prevailing wave direction and therefore experience little or no nett drift. Usually backed by high ground.

(iii) Barrier beaches — sand or shingle beaches backed by low lying ground. May or may not experience a nett sediment drift.

(iv) Duned beaches — sandy beaches backed by a system of dunes.

On each of these beach types sediment will be moved by the action of waves, currents and wind. This transport can be considered to occur in both a longshore and cross-shore sense.

The nett cross-shore transport direction in the surf zone can be either onshore or offshore depending on the incident wave conditions and the beach slope. Generally, on a beach with its profile close to the equilibrium slope, material will be moved offshore during storm conditions allowing a wider surf zone to develop, and onshore under more gentle swell conditions as the surf zone narrows. In its response to waves the beach effectively maintains a uniform dissipation of wave energy per unit volume of surf zone, the surf zone expanding and contracting as incident energy levels rise and fall. Although in the short-term, i.e. over a few hours or days, movement of beach material onshore or offshore can produce significant seaward or landward movements of the shoreline, over a period of years the integrated effect of onshore/offshore transport is usually small. Provided that there is sufficient volume of material within the profile to prevent banks being breached or backing seawalls being reached by waves, the beach will usually return to much the same cross-section and position despite often substantial changes in the intervening months.

On most coasts the major cause of long-term change in the beach plan shape is alongshore sediment transport or, to be more precise, the variation in longshore sediment transport along the coast. If the volume of sediment leaving one section of beach to, say, the east is replaced by an equal volume of sediment arriving from the west, then clearly no change in the cross section of the beach will occur. If, however, the volumes arriving and departing are unequal, then there will be a movement of the beach contours, either landward or seaward, in response to the changing volume of material. The changes in transport rate along a coastline may be due to a number of different factors, either singly or in combination. These may include:

(i) additional inputs or removal of material from the system through either cliff erosion or beach mining,

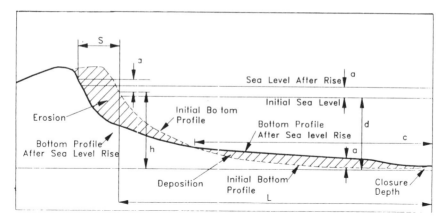

Fig. 2. Representative profile demonstrating Bruun's concept of on/off-shore sediment balancing in response to sea-level rise

(ii) variations in the bathymetry and coastal alignment,

(iii) variations in the incident wave conditions and directions, and

(iv) interference in the coastal regime through the construction of beach control structures such as groynes and breakwaters, or harbour arms.

Although longshore and cross-shore transport processes are common to all beaches, the relative importance of a particular process to the long-term stability of a beach is very much a function of beach type.

2.1. Nett-drift beaches

Nett-drift beaches will be affected by a range of climate change effects. Provided sea level rise is relatively gradual and that there is a sufficient volume of material in the beach to accommodate profile changes, most nett-drift beaches will not be affected seriously by sea level rise. Bruun (Ref. 1) proposed the principle of a beach equilibrium profile in order to facilitate the prediction of shoreline erosion caused by sea level rise. The basic tenet is that for a given wave climate and sediment size a beach will attain a predictable equilibrium profile. As sea level rises material will be eroded from the inshore section of the profile and deposited further offshore, thus maintaining the equilibrium profile relative to the mean water level. This effect is illustrated schematically in Fig. 2. Further discussion on the usage and restrictions of the 'rule' are given in Ref. 2.

A simple assessment of the 'Bruun rule' shows that the relationship

Table 1

Beach Slope	Beach Material	Minimum Recession (m)	Maximum Recession (m)
1:100	Fine sand	14	24
1:50	Medium sand	7	12
1:20	Coarse sand	3	5
1:7	Shingle	1	1.7

between sea level rise and shoreline recession is predominantly a function of beach slope (Ref. 3). Thus for a range of typical UK beach slopes we can make an initial estimate of the potential shoreline recession due to an anticipated sea level rise of between 14 and 24 cm by 2030 (Table 1).

Superimposed on this gradual recession due to sea level rise will be a change in longshore sediment transport patterns due to changes in wind and hence wave direction. These changes may result in increased erosion (or accretion) or may even cause erosion where none existed previously. in the extreme, complete reversals of drift direction may occur such as has already been observed at West Bay, near Bridport, Dorset, and to a lesser extent to the east of Pagham Harbour entrance (Fig. 1).

At West Bay the historic east-to-west drift resulted in a substantial build-up of material against the eastern harbour arm. Since the early 1980s, however, the drift has reversed resulting in the beach receding at rates of up to 2 m per year. This drift reversal appears to be linked directly to a significant reduction in storms from the south-east sector since 1983, and a more recent increase in the frequency of storms from the south-west (Ref. 4). As such, West Bay stands as an example of the problems that may be expected to occur at other locations around the UK coastline under a scenario of continuing climate change.

As drift patterns change, the performance of, and perhaps need for, existing beach control structures (e.g. groynes, detached breakwaters, etc.) will also alter. Old systems may require substantial modification or abandonment, and new systems may need to be built. Many of our current beach management techniques will also need to be re-thought. The need for and extent of re-cycling of sediment from one end (downdrift) of a beach to the other (updrift), such as practised currently at Dungeness and Seaford (Fig. 1), will be affected by variations in the nett drift direction and volumes. The frequency of re-cycling and perhaps the source of material may have to be re-considered. In extreme situations, perhaps even the viability of the operation may be brought into question.

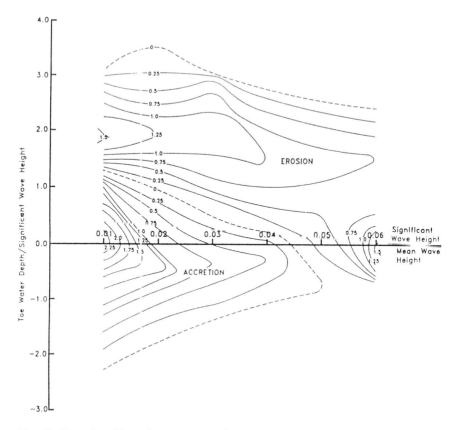

Fig. 3. Trends of beach erosion and accretion in front of a vertical seawall

On top of the longer term changes will also be the short term effects brought about by climate change. As wave heights and storm frequency increase, beaches will have less time to recover between consecutive events. We may therefore expect greater profile variations and, as a consequence, will require increased volumes of material within the profile to accommodate these changes. As a general rule, the minimum beach volume required at a site to provide an acceptable standard of protection will need to increase. If there is no increase there will be a higher risk of the beach being drawn down and of backing hard defences (i.e. seawalls) being subject to wave action. Research has shown that in situations where the beach is sufficiently depleted to allow a backing seawall to influence the surf zone processes, the beach becomes more vulnerable to wave action and experiences erosion over a far wider range of wave conditions than it

Table 2

Increase in average windspeed (%)	Increase in aeolian transport (%)
1	3
5	16
10	33

would if the influence of the seawall was removed (Fig. 3). It is therefore important in situations where the beach is an integral part of the coastal defences to ensure that either

(i) there is a sufficient volume of material within the beach to prevent waves reaching the seawall, or

(ii) the beach is re-instated as quickly as possible following seawall exposure.

One final aspect of climate change that could be of relevance to nett-drift beaches is the potential alteration in aeolian transport due to changes in average windspeed and direction. These changes may result in the increased loss of fine sediments to the hinterland or increased long-shore transport. Hsu (Ref. 5) proposed a method for relating aeolian sand transport to the average windspeed and the sediment characteristics. For a particular beach Hsu concluded that the sand transport, q, (kg/m/s) was proportional to the average windspeed, U^3. Using this relationship, we can make an initial estimate of the potential increase in aeolian sand transport due to an increase in average windspeed occurring as a result of climate change (Table 2).

In practice, in most situations, much of the sand carried landwards by aeolian transport is intercepted and trapped by seawalls and other struc-tures at the rear of the beach. The losses into the hinterland are often, therefore, relatively small and are unlikely to be much influenced by climate change effects. This will not necessarily be the case on duned coasts which are covered later.

2.2. Pocket beaches

Pocket beaches will be subject to a range of climate change effects similar to those affecting nett-drift beaches and in many respects the response will be the same for both beach types. The major difference, however, between a nett-drift beach and a pocket beach is that the pocket beach, by virtue of its orientation into the dominant wave direction, experiences little or no nett drift. This situation will change if the dominant wave direction changes: the beach will no longer be in equilibrium with the wave climate and a nett drift will be induced causing the beach to alter its alignment. Because there is usually a finite volume of material on pocket beaches any longshore migration of material will necessarily result in a thinning, or even loss of the beach over some lengths of the frontage. This may result in previously secure lengths of coastline being exposed to wave action with the consequent risk of erosion and/or flooding.

2.3. Barrier beaches

Most barrier beaches front low lying agricultural land or salt or grazing marshes and provide the primary protection to that land against inundation by seawater. A number of these beaches, such as Porlock and Hurst Castle Spit (Fig. 1), have over the years been severely depleted as a result of man's interference in the natural longshore drift regime. This has left them highly susceptible to sea level rise and associated climate change effects.

In general terms the response of a barrier beach to climate change and sea level rise will be dependent on the balance between sediment supply and erosion. Carter (Ref. 6) identifies three possible situations:

(i) Where sea level rise, or an increase in wave height, promotes an increase in sediment supply in excess of the increase in beach erosion, the barrier will build upwards and seawards. This situation also holds for nett-drift beaches.

(ii) Where the increase in sediment supply is not sufficient to allow the barrier to build seawards but is sufficient to allow it to prograde upwards.

(iii) Where the change in sediment supply is not sufficient to compensate for an increase in beach erosion and the barrier is overtopped (overwashed) and pushed landwards.

It is this last situation which will be of most interest to coastal managers and engineers under a scenario of climate change.

Nicholls and Webber (Ref. 7) identified two distinct forms of overwashing leading to beach recession (Fig. 4):

(i) Crest-maintaining overwashing which occurs without a significant reduction in crest height, and which produces a reasonably steady 'rollover' of the barrier. During this type of overwashing most of the wave uprush is confined to the seaward side of the beach.

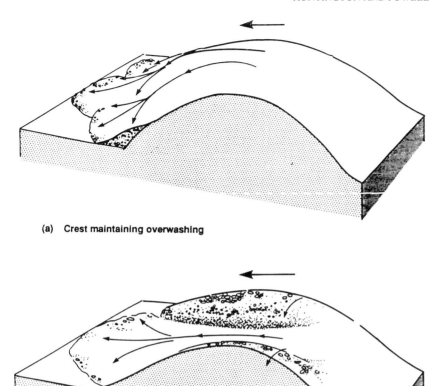

(a) **Crest maintaining overwashing**

(b) **Throat confined overwashing**

Fig. 4. Overwash mechanisms on barrier beaches (after Nicholls and Webber)

(ii) Throat-confined overwashing which reduces the crest height, causing major failure of the barrier beach and allowing substantially shore wave penetration and onshore sediment transport than crest-maintaining overwashing.

An extension of throat-confined overwashing which reduces whole lengths of a barrier to washover flats can occur in extreme events. This process is commonly known as overstepping.

One of the best examples of a beach suffering from these overwash effects is Hurst Castle Spit in Hampshire (Fig. 1). With a declining volume of shingle the spit is a transgressive feature moving landwards due to the processes of overwashing. Up to 1968 the rate of transgression was approximately 1.5 m per year, increasing to 3.5 m per year between 1968 and 1982. Since 1982 the spit has been subject to frequent overwashing and crest lowering during storms. Extensive throat and overwash fan systems have formed leading to a displacement of material into the marshes in the lee of the spit, and a consequent further decline in the volume of the beach above the low water mark. The throat-confined overwash events, which were once localised and sufficiently small to allow the crest to reform naturally, have now grown in frequency and size to such an extent that the beach requires a considerable length of time to effect a natural recovery.

During the severe storms of December 1989 approximately 800 m (Ref. 8) of the bank suffered overtopping. Crest lowering by in excess of 2.5 m and roll back of the seaward and lee toe of the spit by up to 60 m and 80 m respectively, lead to the displacement of more than 100 000 tonnes of shingle overnight. Re-construction of the bank was only achieved by mechanical intervention, and at considerable cost.

Although studies are now underway to ensure the future of Hurst Castle Spit, the experiences at this site are but an example of what may occur at other barrier beach sites around the UK coastline under a scenario of sea level rise and climate change. Given the importance of these defences in protecting low lying land it is clear that considerable effort may have to be devoted to ensuring their future stability. This will be doubly necessary if the overwashing processes are accompanied by an overall reduction in the volume of material entering the system, due to changing drift rates and/or reversals in drift directions.

2.4. Duned beaches

In many respects the response of duned beaches to sea level rise and climatic change will be similar to that of nett-drift beaches. The dunes provide a reservoir of material which may be input to the beach under storm conditions in order to maintain an 'equilibrium' profile (Ref. 1). This material is then returned to the face of the dune system by offshore transport during calmer conditions, or by aeolian transport at lower tidal levels. Although aeolian transport can aid the recovery of dunes following severe storms, wind is also the primary agent of dune erosion and provides the main control on dune elevation (Ref. 9). Around the UK coast dune systems rarely reach more than 30 m in height and are frequently not much more than 15 m high. Right from the onset of dune formation erosion and

Table 3

Increase in windspeed (%)	Reduction in dune height %
1	10
2	20
5	40

accretion occur side by side but, while the dune is growing, vigorous plant growth encourages more sand accumulation than is lost by erosion. As wind speed increases at higher elevations, the critical height to which the dune can grow is reached: desiccation and mechanical damage reduce the capacity for plant growth, and hence the ability of the dune to trap sand. Erosion now counterbalances accretion and the crest level of the dune stabilizes.

If windspeeds increase due to climate change it might be expected that the critical dune elevation will be reduced. Calculations using a simplistic windspeed profile and an assumed dune height of 15 m seem to confirm this effect (Table 3). In practice, however, the processes of dune height reduction will be complex and will involve a wide range of factors.

As an example, we know that vegetation is vital to dune stability, but we do not know the extent to which that vegetation will be affected by climate change, and how that in itself will influence dune growth and decline. It is unlikely that any reduction in dune crest elevation will be as severe as the above table suggests but there can be little doubt that dune systems will suffer as a result of climate change.

During storm surges significant levels of dune erosion can occur. The storm surge of 1953 on the east coast of England resulted in the loss of up to 18 m of dune (Ref. 9). When the sea can regularly attack a coastal dune the waves erode the base and undercut the upper levels of the dune face which subsequently slumps leaving a near vertical scarp. This scarp is too unstable for plants to colonise and will remain an eroding surface. As storm surges increase in frequency and severity under a scenario of climate change, the extent of this erosion will increase while the recovery period between storms shortens. As a consequence the fragile balance between erosion and accretion which exists on this type of coastline may be irrevocably tilted towards the erosive mechanisms.

3. Beach management requirements

Although it is difficult to predict future events and developments it seems unlikely that any radically new methods for combatting coastal erosion will be proposed. Instead our approach to tackling the consequences of sea level rise and climate change will require a far greater emphasis on predicting future coastline changes and evolving a planning/management strategy to cope with them.

Many of our present day modelling techniques could be applied to provide initial estimates of future coastline evolution under a range of climate change scenarios. Indeed similar work has already been undertaken in the Netherlands, albeit on a much simpler coastline. The modelling would need to consider the implications of changes in drift rates and drift patterns. it would need to recognise that if storms become more frequent and more severe then the volume of material within a beach profile will have to be increased if the overall standard of defence is not to be lowered. The consequences of sea level rise would have to be taken into account either by the application of sophisticated computer models or, perhaps, initially by the use of the simple process models proposed by Bruun (Ref. 1), Hands (Ref. 3) and fellow workers.

Any modelling of coastline changes would have to be backed up by a comprehensive programme of regular field monitoring covering not just waves and water levels but also beach changes, cliff erosion and a range of other environmental parameters. Details of some of the monitoring techniques that will be required are given in Ref. 10.

Once an understanding of potential coastal changes has been obtained, the management responses can be developed. Some systems may have to be left to evolve naturally if modelling has shown that this is the most cost-effective option. For other sites it may be more appropriate to replenish the beaches using material taken from either land-based or offshore sources, or perhaps material recycled from elsewhere on the coast. In an ideal situation the replenished beaches would be open, natural systems, uncluttered by other structures. However, it will always be necessary to balance capital and maintenance costs, and in order to ensure an acceptable maintenance schedule some beach control structures (groynes, breakwaters, etc.) will often be required. In the extreme situation it may be necessary to abandon existing soft defences and to construct new seawalls if the predictive modelling indicates that this is the only way to avoid serious loss of land and property.

The future is not, however, totally bleak, as beaches that are eroded in one area many build up again on an adjacent stretch of coast. This could allow beach based coastal defences to be developed in areas where, previously, they did not exist. Also the capacities of plant to move material will increase with time, bringing about reductions in costs. This, coupled with a greater reliance on dredgers for beach maintenance, will improve our ability to cope with many of the consequences of climate change.

The financial implications of undertaking beach replenishments under a scenario of climate change have been considered by Weggel (Ref. 11) who proposed the use of a 'present worth' factor, P, for evaluating the cost of a beach replenishment project, where

$$P = \frac{(1+i)^N}{(1+i)^N - 1} \cdot \left[1 - \frac{1}{(1+i)^L} \right]$$

and N is the number of years between maintenance replenishments, i is the prevailing annual interest rate and L is the design life of the scheme in years. Weggel also highlighted the fact that sea level rise predictions such as those proposed by Hoffman (Ref. 12) show a projected acceleration in sea level rise rates. The consequence of this acceleration will be either an increase in the frequency of periodic beach replenishments in order to maintain acceptable defence standards, or an increase in the initial size of the scheme, and thereby an extended maintenance interval.

As an example of this effect we can consider the increase in costs for a beach replenishment scheme designed for a 50 year life with 10 yearly maintenance replenishments. If the annual interest rate is assumed to be 10% then the costs of a scheme designed on the basis of an accelerated sea level rise in accordance with Hoffmans (Ref. 12) conservative estimate, and consistent with the scenario presently assumed for the UK, are 15% higher than those for the same scheme designed for a constant rate of sea level rise.

Weggel (Ref. 11) also makes the point that although the costs of beach replenishments may escalate if there is an acceleration in the rates of sea level rise, there is unlikely to be a similar increase in the benefits accruing from these projects. Cost-benefit requirements for beach replenishment schemes may therefore become increasingly difficult to achieve and, unless the funding rules for coast defence works are changed, the balance may once more tilt away from soft defences in favour of the construction of new seawalls.

Ultimately we must recognise that while the extent and direction of climate change is uncertain the potential response of natural systems, such as the coastline, to that change is equally unclear. It is only by attempting to predict in detail the likely response under a wide range of scenario's that we can begin to comprehend the magnitude and cost of the management task before us, and then establish the planning mechanisms that will be necessary to achieve it.

Acknowledgements

The authors are grateful to a number of their colleagues at HR Wallingford for their advice and helpful discussions during the writing of this paper. Particular thanks should go to Dr Alan Brampton and Mr Tony Diserens who jointly developed many of the ideas regarding the extent of climate change on the coastal zone whilst preparing a parallel research report (Ref. 13).

References

1. Bruun P. Sea level rise as a cause of shore erosion. ASCE *Journal of Waterways and Harbours Division*, 1962, **88**, 117 – 139.
2. Bruun P. Review of conditions for uses of the Bruun rule of erosion. *Coastal Engineering*, 1983, 7, 77 – 89.
3. Hands E.B. The Great Lakes as a test model for profile response to sea-level changes. *CRC Handbook of coastal processes and erosion*, Boca Raton FL: CRC Press, 1983.
4. HR Wallingford. *West Bay harbour: Analysis of recent beach changes east of the harbour*. HR Report No. EX 2272, Jan. 1991.
5. Hsu S.A. Computing aeolian sand transport from routine weather data. *Int. Coastal Engng Conf.* 1974.
6. Carter R.W.G. *Coastal environments. An introduction to the physical, ecological and cultural systems of coastlines*. Academic Press, 1988.
7. Nicholls R.J. and Webber N.B. Characteristics of shingle beaches with reference to Christchurch Bay, S. England. *Proc. 21st Int. Coastal Engng Conf.* New York, 1989.
8. Wright D.J. Practical problems of finding suitable materials for beach recharge. *MAFF Conf. of River and Coastal Engrs.* Loughborough, July 1992.
9. Ranwell D.S. and Boar R. *Coast dune management guide*. Inst. of Terrestrial Ecology, Huntingdon, 1986.
10. Brampton A.H. *Coastline Monitoring*. HR Wallingford Report No. IT 345. February 1990.
11. Weggel J.R. Economics of beach nourishment under a scenario of rising sea level. ASCE *Waterway, Port, Coastal and Ocean Engineering*, **112**, No. 3, May 1986.
12. Hoffman J.S. Estimates of future sea level rise, in *Greenhouse Effect and Sea Level Rise*, MC Barth & JG Titus, Eds, van Nostrand Reinhold Co., New York, 1984.
13. Brampton A.H. and Diserens A.P. Beach development due to climate change. NRA project report 281/1/3 (in preparation).

Discussion

Raporteur: M. BARRETT

In the opening of the discussion it was accepted that the uncertainty of the effects of climate change were high and the risks low, while on the other hand protection against such risks may lead to very expensive engineering works. There may therefore be a tendency just to monitor or to do nothing at all about it. Somewhere in the analysis there should be included the cost of the failure occurring. Perhaps there should be an equation which says that low risk or high uncertainty times the cost of the disaster must equal the risk protection costs times some factor. Was this an approach that was used and, if so, what factors were suggested?

Concern was also expressed regarding the graph of predicted SLR which showed a gradual change that will steepen. As there was a lot of 'noise' on this graph, when would one recognise this steepening? Could one afford to wait to undertake remedial works which will be required to prevent future disasters similar to that in Towyn? While not to overdesign was perhaps to design for the engineering optimum, the consequences in terms of public perception of any major failure would be dramatic in terms of civil engineers' reputation.

In responding it was already agreed that monitoring was essential and clearly in any new schemes the design must be tested in terms of predicted SLR and allowances made. There is, however, no way at the moment to make the large investment that would be necessary to maintain the current standards of protection against the rises now being forecast in such a climate of uncertainty. In the Anglian Region of NRA, for instance, there are some 1500 km of sea and tidal defences of which only a few km are replaced each year. The design of the latter is checked very carefully but to raise all 1500 km would be extremely costly. Benefit/cost ratios for any scheme are looked at very carefully and the costs and benefits examined for various solutions with different ranges of flood return period such as 1/100 and 1/1000. In the case of the Colne Barrier a return period of 1/1000 was found to be the optimum.

It was further pointed out that current designs do indeed make allowance for SLR and not just for maintaining the status quo. Historically up to 4 mm per year rise has occurred on the east coast and the current design standard allows for up to 6 mm/a rise over the lifetime of the scheme. In the context of the SLR scenarios quoted this allowance takes one ahead of the predicted rise for some 25 to 30 years. Bearing in mind that the economic life of a scheme may be of the order of 50 years, this design standard actually takes one well into the life of the scheme before some further action might be necessary. After 20 years sufficient data should be available to determine the actual trends. The important thing is to provide designs which are flexible and can be modified if events change.

On the question of bringing uncertainty into the equation, it was not normally brought into the design as such. In the case of the Colne Barrier the design was based on the best estimate of SLR in the area at 5 mm/a and then checked for an SLR of 10 mm/a to see what would happen if the original estimate was found to be wrong and what could be done to accommodate the rise. The barrier was still found to be the best solution and was designed so that it could be raised if necessary, i.e. it complied with a 'no regrets' policy.

An earlier speaker was nervous about the assumption of increased storminess. Whilst accepting that it had been mentioned in the scenario paper he was not aware of any evidence from modelling studies to support this. He had no confidence for giving any signal at this stage of any change in the storminess especially between south and north Britain. Different models had given different results and any apparent trends were within the range of natural variability.

In the context of port engineering a delegate pointed out that climate change is only one of many changes which influence design, such as economics, cargo handling techniques, commodity movements and ship design, none of which can be predicted far ahead. One therefore builds for flexibility, allows for change and accommodates it. To build for certainty in 50 years time is a nice idea but one would soon be out of a job.

In response to the point on storminess it was emphasised that, in the short term, a potential increase in storminess was far more important than SLR in its effect on beaches. In the absence of any confidence in current predictions one must be cautious and consider the possibility of an increase.

Some 25 000 t of beach feeding at Dungeness for a period of over 20 years was mentioned. The question was put as to whether this was a sensible way of proceeding and whether there were other equivalent examples or proof to show that it actually worked. Could the Dungeness results be useful in a general way elsewhere?

The method used at Dungeness was in fact 'recycling'. The operation had been revalued every few years and was still considered to be the best and cheapest way of protecting the power station. There were examples elsewhere along the south coast where this process was used by the NRA. In cases of strong littoral drift where it was known exactly where beach material was coming from and going to it was considered that the benefits of recycling could not be disputed.

Renourishment on the other hand had been used in two ways. It had been used on an emergency basis when excessive material had been lost under storm conditions. Beaches come and go and perhaps such expenditure might be used more economically if one knew what was going to happen. However, where the situation can be modelled in advance and appropriate measures are taken to retain the material then renourishment can be a satisfactory and economical way of carrying out sea defences.

The earlier comments of one delegate regarding design uncertainties were endorsed. These can include weather conditions, extrapolation of extreme events on inadequate data and estimation of extreme wave heights which might be 20% to 30% out. The whole basis of design is uncertainty with or without climate change. We should therefore be using sensitivity tests and bringing a probabilistic approach into design to see what might happen if we have got things wrong. Apropos the reliability of beach feeding, there has been extensive experience abroad, including Europe, USA and Australia, where it has been used with success, particularly on sand beaches. This has shown that, provided it is understood what is going on and conditions are monitored, the chances of a successful scheme are good.

It was asked whether the NRA have a short term strategy for dealing with emergencies in a similar way to the marshalling of highway plant to deal with snowfall and, on another topic, whether fuel technology was moving in the direction of reducing or absorbing CO_2 emissions.

It was replied that the NRA have a proper warning procedure across their whole area. An east coast flood warning procedure comes into play and a model is run by the Meteorological Office giving early warning if a surge is likely. This provides considerable detailed information about flood risk so that services can be mobilised accordingly. Interestingly, on 23 September 1992 this service was actually used, albeit for a fluvial flood, so it has been recently tested in a real event as well as in regular emergency exercises.

On the issue of CO_2, the emission from methane is half that from coal, while the nuclear alternative avoids it, but these would only appear to be options for the developed world at the present time.

In conclusion there was some concern from the opening of the discussion that engineers were not thought to be taking climate change seriously. The meeting had heard how we were learning how to model response to climate change and to monitor conditions. However, it was not just a question of wait and see. Decisions were being taken now to use design options which are less sensitive to possible effects of climate change than others.

Geotechnics and structures in the uncertainty of climatic change

J.R.A. GAMMON, Dar Al-Handasah Consultants

1. Introduction

The purpose of this paper is to identify the factors that influence geotechnical and structural engineering and to examine these in the context of climatic uncertainty. The geotechnical factors range in their time-scale and scope from geological phenomena through to environmental protection and the structural factors include on-shore and off-shore conditions. The processes of investigation, interpretation, design, specification, construction and monitoring receive attention. The interaction between ground and structure — the dependence of one on the integrity of the other— is widely recognised in terms of both temporary and permanent works. In this paper the influence of climate on the in situ behaviour of the ground as well as its extraction for use in a structural sense — as fill or as aggregate — forms part of a broader appreciation of the interaction involved.

The types of project examined vary from large embankments to deep excavations and from highway pavements to high-rise buildings.

2. Geology and climate

2.1. General

Climatic extremes are often closely allied with the changes in types of rock and soil that form the geology of a particular country or continent. The older rocks are frequently a product of an environment alien to humans. Climate viewed on that time-scale has tended to be the subject of speculation, some of which has been satisfied in recent times by the exploration of distant planets where climates different to those existing today on Earth can be studied.

2.2. Ice ages

The most recent long-term climatic change (but short in geological terms) has been associated with the Pleistocene ice ages which started about 1.5 million years ago. The last ice sheet to extend into southern England receded about 11 000 years ago. To place the time-frame of the ice ages into context, it should be noted that the glacial period can be split into two parts separated by the Great Interglacial which alone has been estimated to have lasted about 200 000 years.

An average global temperature change of about 6°C represented the difference between the warm and cold periods during the ice ages.

Average annual temperatures continued to rise after the glacial retreat and are estimated to have been at their highest about 5000 years ago when extensive deciduous forests covered much of Great Britain (ref. 4). Higher temperatures than at present also occurred during the Medieval warm period between 600 to 900 years ago. However, overall temperatures have tended to fall and it is against the potential for a significant reversal of that trend — as opposed to short-term 'interference' — that the present day concerns for climatic change have to be set.

As the ice ages are the most recent extremes in environmental conditions and with ice covering about one-third of the Earth's land surface, they have been responsible for many of the present-day landforms. The mechanisms involved range from the direct gouging of valleys to the large-scale uplift by hundreds of metres of areas of ground relieved of the burden of ice which measured in some instances over 1000 m in thickness. Such isostatic recovery is still in progress and unlike other extremes in the geological past it is also still possible to see glaciers in action. The behaviour of the Fox and Franz Josef glaciers on the west coast of the South Island in New Zealand provides an opportunity to see the advance and retreat of a glacier over a period of decades instead of centuries or millennia.

2.3. Periglacial phenomena

Ground conditions in areas bordering the glaciers of the ice ages but not directly loaded by ice have generated their own engineering problems and the detection of periglacial phenomena is extremely important (ref. 23). Present day examples of similar conditions can be found in the tundra areas of the North American continent (ref. 6).

2.4. Sea level changes

During the ice ages sea levels rose and fell dramatically in relation to present day conditions. With large quantities of water held in solid form as ice the resultant lowering of sea levels, by more than 100 m (ref. 29) was accompanied by a lowering of valley floors through erosion by downward-cutting rivers. Sea level rises led to the infilling of the valleys with great thicknesses of sediments in some instances. Rocks were exposed to extremes of weathering during these fluctuations (ref. 5).

2.5. Present climatic time-frame

An important feature of the climatic changes associated with the ice ages and which enhances their relevance to the present day potential for climatic change is that they were contemporaneous and world-wide in their effects (ref. 29). The climatic time-frame relevant to today's engineering is far shorter than that associated with geological changes. With the potential for climatic change related to changes in the concentrations of gases in the atmosphere, particularly carbon dioxide, the relevant time-frame is that associated with the rise in industrial activity and the ensuing period to the present day. This essentially means the period from about 1870 to today, and detailed studies of the climate are normally contained within this time-frame. Projections of climatic change into the future are related to the time at which the proportion of carbon dioxide in the atmosphere is expected to double in relation to pre-1870 conditions; this normally sets the projections as far forward as between 2050 to 2100.

The time-frame for climatic change is therefore in the realms of the 'foreseeable future'. The debate about the significance or otherwise of climatic change — and the validity of classing the observations already made as change on the climatic scale — can be followed in the letters columns and technical papers in journals from countries as diverse as the United States of America (ref. 7) and New Zealand (refs 8, 15) as well as the United Kingdom.

3. The geotechnical time-frame

Geologists, engineering geologists and geotechnical engineers are all fortunate as their training in, and experience of, geology enables them to work comfortably in the time-frames over which significant climatic change might be expected to occur. A grasp of both chemistry and physics and the importance of the chemical and physical environments that naturally occur over millennia also help in this regard. Such people working in civil engineering and with the diversity and uncertainties of natural materials are trained in observation and prediction but do not expect always to be correct in their hypotheses.

4. The structural time-frame

4.1. The concept of design life

In contrast to considerations of the ground itself, the structures resting on or lying within the ground are normally contemplated by their designers in terms of 'design lives', often shorter in duration than the average human life expectancy and hence less vulnerable — at first glance — to the

127

time-scale of significant climatic change. However, this does not permit complacency as society's dependence on the integrity of certain structures grows with time. Thus bridges are designed for significantly longer lives, usually 120 years, when compared to buildings which are usually designed for 50 years. Some structures in harsh environments have performed extremely well; John Smeaton's pier at Ramsgate, now well over 200 years old, is a particularly good example (despite the need in recent years for remedial work).

4.2. Ancient versus modern

The robustness and durability of ancient structures, which have lasted over a thousand years in different types of climate around the world, is worthy of examination. However, present-day economic constraints place pressures on structural designers to utilise structural materials more and more efficiently and to build more rapidly (ref. 1). Once the optimum structural solution to loadings prescribed by the codes is still found to be too expensive, then the loadings themselves come under scrutiny. This is considered a worrying development in that the apparently small reductions in floor loadings being sought by some developers (ref. 2) are being viewed in isolation of the increases in load or changes in type of load that climatic change could bring about — and bring about well within the actual life of buildings as opposed to their (usually) shorter design life. The uncertainty of climatic change predictions introduces yet another area of design speculation but one that most structural engineers appear to prefer to avoid. The generally cautious nature of such engineering is not compatible with unproven hypotheses or the contemplation of predictions being wrong.

4.3. Structural form in the climatic time-frame

The time-frame over which some of the structural forms have been in existence (e.g. reinforced concrete and prestressed concrete) is relatively short — certainly in terms of the time over which humans have been able to monitor climate. Some of the problems which have occurred can be seen to be related to the environment in which the structures are located but in many instances structures have not experienced the extreme conditions anticipated from observations of the climate to date as opposed to the consequences of predicted changes in that climate. Furthermore, the wind responsible for the destruction of the Tacoma Narrows bridge (ref. 3) was not exceptional and environmental loading resulting in other failures has not always been extreme (ref. 24).

4.4. Durability

The need to guard against corrosion or other forms of physical or chemical degradation, particularly where structural members or anchors are in tension, is strongly related to the climate of the area in which the structure is located.

Concerns for the durability of 'new' materials such as geosynthetics (geotextiles, geomembranes, etc.) appear disproportionate when viewed against problems encountered with the basic materials of construction such as concrete, steel, and timber when placed in environments already associated with the present-day climate. However, geosynthetics suffer from relatively brief observation of their working behaviour and demands are made for proof of their long-term integrity as opposed to the manufacturer's predictions (ref. 26).

Restoration work on the fabric of old buildings is primarily due to pollution. The two factors of an increase in carbon dioxide and pollution are related (through increased industrial activity) but must not be confused when assessing the effects of climatic change alone.

The time-frame for the observation of structural behaviour using rapidly advancing technology has been short. If climatic change predictions are correct and where these lead to more onerous loading conditions or more onerous conditions in terms of durability, then such predictions deserve more than passing thought.

5. Climate modelling and predictions

5.1. Prediction and proof

Predictions of climatic change are made on the basis of complex mathematical models that can be analysed using computer techniques. These programs are similar, in terms of handling variables in a complex natural system, to those used to predict the movement of fluids in the ground or the occurrence of hydrocarbons such as oil. Geologists and geotechnical engineers should be familiar with these techniques and the shortcomings associated with them. However, this recognition of shortcomings does not stop such programs being used. Very expensive oil exploration drilling is undertaken on the basis of prediction that cannot yet — despite the apparent wealth of data and previous experience of drilling successfully elsewhere — be described as certain. It has been argued (ref. 9) that those already familiar with an absence of certainties before embarking on a task, such as occurs in geotechnics, should not be demanding proof before taking actions associated with climatic change.

5.2. Global effects and carbon budget

The models are measuring sensitivity such as the effect of doubling the carbon dioxide concentration on global temperatures. In trying to complete the prediction in terms of time it is necessary to predict the time taken to bring about this doubling of concentration. This is extremely difficult. The relationship between the systems producing carbon dioxide and the amount encountered in the atmosphere constitute the 'carbon budget'. The difficulties associated with time-related aspects of the prediction are complicated by a significant percentage of carbon dioxide production that cannot be accounted for in the atmosphere (20%, ref. 15) and uncertainty about the way biological and ocean systems will in fact respond.

5.3. Regional effects

As with all modelling of this kind the size of grid cells is important. In order to keep the task to manageable proportions grid sizes have tended to be large; e.g. 1000 km (ref. 17). Thus variability within regions proves difficult to predict with the same degree of certainty as the global predictions. However, the situation here is further complicated by the influence of local features such as mountains, particularly where these occur at the coastline. Climatic change will not 'flatten' these features and they will continue to produce extreme conditions over short distances as at present. New Zealand's Fiordland, towards the southern end of the west coast of the South Island, generates an annual rainfall of over 8 m from the moisture laden air coming from the Tasman Sea, yet a few tens of kilometres inland the Otago Plains suffer prolonged periods of drought. This juxtaposition of storminess and drought, which occurs in a moderated form in the United Kingdom, should not be expected to change dramatically as result of climatic change.

Studies show the importance of regional climatic and physiographic characteristics to regional climatic change (e.g. ref. 14). Comparisons between hydrological effects in similar topography but subject to different climates are valuable. Such studies can indicate changes in surface runoff and soil moisture which have engineering significance. It may be found that snow, accumulating and melting, is the most significant parameter in basin response to climatic change and this will depend on altitude–area–elevation relationships and orientation (ref. 14). As a further example of regional effects, it can be noted that parts of Canada and Greenland occupy the same latitudes. During the ice ages both were covered in ice. Canada now has relatively little precipitation and glaciation compared to Greenland. This is due to the Atlantic Ocean bordering Greenland and supplying sufficient moisture to sustain its present state of glaciation. The smaller the country the more difficult it becomes to predict the changes which would significantly affect local engineering practice.

Sea level rises, however, are almost immune to the regional uncertainty that affects wind and precipitation predictions and it is this consequence of climatic change which receives detailed attention in other papers in this symposium. It is also this type of change which could significantly modify the conditions at coastline mountain ranges; river levels and discharge capacity at the mouth of rivers then need to be examined more carefully and also in terms of their upstream effects which could include flooding and failure of river banks (ref.17).

6. The uncertain factors

The assessment of climatic change involves examination of the changes in air, temperature and water. These factors not only influence but are also influenced by climatic change. They are interrelated in a complex and dynamic way.

6.1. Air

The importance of air is that it is the changes in the gases making up the air which are seen as the reason for the climatic changes in temperature and precipitation. Modelling of climatic change normally adopts the premise of a doubling of the carbon dioxide content in the air. The significance of carbon dioxide is that it contributes more to the enhanced greenhouse effect than all the other so-called human-derived gases such as methane, nitrous oxides, chlorofluorocarbons and ozone (ref. 9). It is the vast volume of carbon dioxide that causes problems, as its weight-for-weight potency is far less than these other gases.

In terms of agencies in the atmosphere responsible for engineering problems these gases do not figure prominently. The attempts being made to cleanse the air of pollutants have generated significant work of a positive nature for engineers. Where fluctuations in the atmosphere lead to increases in the ultraviolet light reaching the Earth's surface then some construction products are put at risk; some geosynthetics and some types of pipework, for example, need to be protected against ultraviolet degradation. In view of its significance to corrosion, either directly or as a medium for transporting aggressive chemicals through the air, water vapour is a very important constituent in the atmosphere in engineering terms. It also has the largest greenhouse effect but loses its significance in terms of being relatively unaffected by human activity (ref. 10) in contrast to carbon dioxide and hence little can be done to moderate the effects of water vapour in the greenhouse equation.

It is when air moves from one place to another that it becomes of significance to structural and geotechnical engineering. Wind loading has to be carefully evaluated and can represent a major design constraint for high rise buildings. Building codes tend to develop along national lines, reflecting the scale of construction undertaken as well as the measured and statistically predicted wind strengths. Wind generates lateral forces on buildings which need to be adequately accommodated by the foundations as well as the structural frame (ref. 24). In Hong Kong the high cost of land has meant that tall buildings are constructed in a region frequently struck by typhoons and their associated wind forces.

Reference has already been made to the significance of the ice ages in climatic and engineering terms. The ice formed at that time not only gave tangible evidence of climatic change it also managed to record the condition of the atmosphere through the entrapment of air bubbles. By analysing these air samples (ref. 11) it has been found that the maximum natural carbon dioxide level during the interglacials amounted to about 280 parts per million by volume (ppmv). Currently levels are at about 350 ppmv with about 50% of the rise occurring since 1960.

Direct barometric pressure effects in addition to the generation of wind should be examined. It has been found that changes in barometric pressure can have a significant effect on groundwater regimes as well as open bodies of water. Links between barometric pressure, seismicity and volcanic activity warrant further study in order to assess further the significance of climatic change.

6.2. Temperature

The detection of gases in the air, of significance to climatic change, is beyond human perception. The ability to monitor gases and isolate them for their concentrations is a recent development. Although individuals demonstrate different tolerance to changes in temperature, the perception of such changes is a human characteristic. The measurement of temperature is well-established and the development of long-term reliable records of sufficient sensitivity strongly parallels the period over which carbon dioxide concentrations have increased.

Temperature changes influence the state and mode of transport of water (as rain or snow, as rivers or glaciers, for example) and differential temperatures cause the movement of air and hence create winds.

Past records tend to fall into blocks of time. For example, data reported for northern latitudes — from 64° north to 90° north — indicates that during 1800 to 1940 annual average temperatures rose by 2°C, from 1940 to 1970 temperatures fell by 1°C, and records showed an increase of 0.3°C from 1970 to 1980 (ref. 6). In contrast, the relatively localised area of the northern slopes of Alaska experienced a net increase in temperature of between 2°C and 4°C during the period 1900 to 1990. Additional data can be found in the other papers in this volume.

Predictions of temperature change (e.g. refs 6, 12, 14) usually show these annual average temperature changes in global and regional terms based on a doubling of carbon dioxide concentration. The significance of this is that although there is a growing consensus regarding the magnitude of the global annual average — an increase of between 1.5°C and 5°C — the regional averages show ranges such as –3°C to +10°C. Some texts give a single value for the global average (e.g. +3.7°C by the year 2100, ref. 18; 1°C in the next forty years and 3°C by 2100, ref. 9). A trend in changes in global temperatures has also been related to latitude, with the increase in temperature expected to be higher at increasing latitudes. Towards the poles average temperature increases could be between three to five times the global mean.

The temperature variations described should also be considered in the context of a total absence of greenhouse effect which, it has been estimated, would reduce the average global temperature by 33°C (ref. 17).

As with the behaviour of winds, the extremes of temperature can be observed on Earth at the present time. It is not as though climatic change would uniquely create deserts or uniquely create frozen masses of water. However, such extremes are only tolerable for long periods if an artificial environment for living purposes is created. Moving away from these extremes, the requirement for comfort in the working environment has led to the increased adoption of air-conditioning in buildings. These situations have particular significance for structural engineering in terms of increased mechanical and electrical equipment loadings and the accommodation of additional ducting, for example.

6.3. Water

As well as being vital to life, water is a major factor in many engineering works. Although its consumption can be tolerated by humans over a wide range of temperatures, the environments associated with its natural occurrence beyond these temperatures (i.e. towards freezing in arctic regions and towards boiling in geothermally active areas) demand special measures to be taken in geotechnical and structural engineering.

The long-term recording of temperature has been noted above. The incidence of water, particularly where it occurs as precipitation (rain, snow, hail, etc.), has also formed the subject of extensive records, available on a world-wide basis. Where both water and temperature are important, as they are in agriculture and horticulture, then particularly extensive and reliable records can be found. Meteorological stations are established on a scientific basis to monitor as many phenomena as possible, and rates of evaporation and humidity may also form part of the data available. The direct temperature of large bodies of water, such as the sea, as opposed to ambient temperatures during precipitation, is significant as sea temperatures rising to about 27°C are associated with the onset of typhoons in areas such as the South China Sea.

High temperatures and an absence of water creates problems too, not just in terms of a threat to life, but also where soils shrink due to drying out, and subsidence occurs.

The incidence of water, in any of its states (including water vapour) is therefore very important and the effects of the temperature changes described in the previous section of this paper are also translated into predictions of changes in water-related phenomenon; e.g. precipitation, areas of land covered with ice, changes in the thickness of frozen ground (tundra), changes in soil moisture deficit for irrigation, increases in sea-level, etc. The most readily measured phenomenon, available on a global basis, is precipitation.

As with temperature, changes in precipitation associated with climatic change are provided in terms of global and regional averages. A distinction should be drawn for engineering purposes between the proportions of the change in precipitation occurring as water or as snow (ref. 14). It has been predicted that global mean precipitation could increase by 7% to 15%. This initially appears encouraging for areas currently experiencing droughts. However, as can be found in other papers in this volume, the regional averages for precipitation show a range from a decrease by 25% to an increase by 25%. Part of this regional variation can be attributed to the ability of a warmer atmosphere to retain more water vapour. Global warming can be expected to intensify the hydrological cycle in coastal areas and over open ocean, hence its significance to structural engineering in these environments, but the interior areas are generally expected to be drier (refs 13, 17) which may prove of more significance to geotechnical engineering where soils affected by changes in moisture content occur. Increased amounts of water vapour could also result in an increase in the amount of snow falling.

Sea level rises, for example due to the expansion of warmer seawater and the melting of mountain glaciers, of up to lm in the next hundred years have been predicted (ref.17, 25).

Irrigation studies undertaken in terms of climatic change are important because they identify the potential need for significant engineering works to convey water to where it is required. Indeed, the problem with water is that it is not readily available when and where needed (ref.12); the actual amount of fresh water needed to sustain life globally represents a small fraction of that available from rivers alone. Irrigation studies also attempt to distinguish between temperature-only change and temperature-and-rainfall change in terms of water budgeting and these two situations can also be contrasted for their engineering significance.

7. Investigation

7.1. General scope

In geotechnical studies the process of investigation usually involves a 'desk' study, fieldwork, laboratory testing, and monitoring. The object of the investigation is to establish the ground conditions and groundwater regime of significance to the proposed engineering works and to provide factual data from which engineering parameters can be interpreted for design and construction purposes. Climatic change is not expected to change this objective but rather to place particular emphasis on the need to obtain data from long-term monitoring, for example, and to examine in more detail aspects of the soils and rocks such as their chemistry and their susceptibility to change due to temperature variation and the addition or removal of water.

7.2. Ground conditions and investigation techniques

Ground conditions are usually examined in terms of materials left in situ and those to be extracted for incorporation into the Works, as fill or concrete aggregate for example. Ground investigation techniques are continually evolving, particularly in the area of in situ testing, and are described in a wide range of textbooks, journals, and specifications (e.g. refs 30, 31).

The engineering behaviour of the ground is related to the state and chemistry of the water within the soils and rocks and the ability of water to move through the mineralogical skeleton, due to the size of the particles and the bonding, if any, at their contacts.

7.3. Particle size distinctions

Geotechnics attaches great significance to the particle sizes associated with soil and rock. Clays lie at the smaller end of the soil particle range where particle mineralogy is significant and chemical bonding is predominant. Pore sizes are correspondingly small and the permeability is very low. Soils are usually classified as clays when the particle sizes are less than 200 microns (0.002 mm). At particle sizes larger than 0.06 mm (fine sand and coarser material) the chemical bonding is relatively insignificant and permeabilities become much higher. The significance of permeability lies not only in the transmission of water under natural conditions but also the rate at which water can be displaced or the way in which pore pressures can change when a soil is subject to physical change (e.g. applied load or temperature).

Silts occupy the particle size range between clays and sands and often exhibit enigmatic behaviour which requires very careful interpretation.

As a generalisation the soils and rocks formed from the smallest particle sizes, such as clays and mudstones, are more vulnerable to the effects of climatic change. Water level or pressure changes can, however, affect the engineering behaviour of all types of soil and rock.

7.4. Surface water and groundwater regimes

The detection and monitoring of water as precipitation, surface water and groundwater forms an important part of the investigation process. The presence of water of engineering significance is affected by climatic change in many different ways. For example, changes in precipitation affect river levels, surface erosion and the growth of vegetation, groundwater levels (short- and long-term) and aquifer recharge (both short- and long-term). Changes in temperature, whilst also affecting the nature of the precipitation (i.e. as water or snow) also influences factors such as evaporation, evapo-transpiration, water–ice balances (in tundra), and the weakening of rock masses due to expansion and contraction.

Both precipitation and temperature have the ability to change the moisture content of soils, which is of great significance to strength for the finer-grained soils either when the soils are left in situ or when used in earthworks.

An indirect effect of climatic change, particularly where temperatures rise, is the change in water demand for uses such as human consumption, irrigation, and air-conditioning systems. The increased abstraction of water can generate engineering problems such as large-scale settlement; increased demands are also placed on disposal systems. Thus groundwater changes in urban areas require special attention (ref. 28).

Surface water can cause direct erosion of the ground surface and generate engineering problems in this way. Where soils are particularly susceptible to erosion (usually where silts and sands predominate) then rainfall patterns need to be studied for their frequency and intensity. In assessing the consequences of the increases in precipitation described in previous sections of this paper, it would then be important to know if the increase occurs as a greater frequency of previous storm events or is due to appreciably higher intensities or durations associated with the same number of storms as have occurred previously.

Interaction between surface water and groundwater in terms of infiltration also needs to be investigated. Where a permeable material overlies a relatively impermeable material then there is the potential for the infiltrating water to 'perch' on the boundary between the two; the rise and fall in the perched water table can be rapid and related to the duration of storms if these are responsible for the infiltration on an intermittent basis. The significance of this is that if the boundary is sloping and is reflected by the topography then conditions leading to slope failure can occur. Where a potential problem of this kind is identified then it will be necessary to have continuous monitoring of the groundwater regime. Intermittent readings may miss the time when the perched water table is at its highest and a false indication of the amount of fluctuation may be gained. As an example, in Hong Kong transient rises in the water table of the order of 2 m to 5 m above normal water levels are observed in association with storm events and significant antecedent rainfall.

7.5. Laboratory testing

Laboratory testing is usually directed towards the classification of soil and rock, strength and compressibility, compaction or earthworks related behaviour (including frost susceptibility where appropriate), and the chemistry of soil, rock, and groundwater.

Fortunately the classification tests (especially particle size and Atterberg limit tests) together with natural moisture content data provide a rapid and economical means of evaluating the susceptibility of materials to the consequences of climatic change which may have been identified.

7.6. Chemical considerations

Chemical changes in soils due to leaching out of salts can be dramatic. In Scandinavia the depositional environment for some clays was saline. Due to a combination of sea level changes and isostasy, the salinity of the pore water was reduced due to leaching by freshwater infiltration. This dramatically reduced the strength of the clay when disturbed (ref. 19). Higher temperatures can lead to higher evaporation loss and this can draw concentrations of chemicals towards the ground surface. Climate-induced chemical change can therefore be very significant (ref. 20). The chemistry of the ground not only affects its own behaviour but also has the potential to affect the durability and strength of structures in contact with the ground (ref. 22).

Changes in chemistry and related effects are not limited to soil. Degradation of cut slopes in granite due to rapid weathering has been noted in Hong Kong, for example, where hot and humid conditions occur. The influence of climate on rock quality has been examined in terms of working with the rock as encountered in situ (ref. 5) and in terms of its extraction for use as construction material (ref. 21); climatic conditions which lead to either disintegration, due to physical weathering, or decomposition, due to chemical weathering, can be distinguished using characteristics based on free-water evaporation and total annual precipitation, for example.

7.7. Monitoring

Monitoring of conditions prior to, during and after construction needs to considered an integral part of the investigation process. Monitoring needs to be accurate, with a high degree of reliability to detect change.

Structural engineering concerns would require the monitoring of wind speed and direction, temperature variation and humidity. Fog detection may also be required. Other monitoring associated with hazard avoidance may need to be carried out as a consequence of climatic change (ref. 24).

8. Interpretation and engineering properties

An optimum site investigation can only be based on a good desk study of the existing available data. Similarly the successful interpretation of factual data to determine engineering properties and the design parameters depends on reliable fieldwork, laboratory testing and monitoring. With clays, susceptibility to swelling or shrinkage due to the addition or removal of water, respectively, under conditions of climatic change, is particularly significant. Swelling causes a decrease in strength or generates high pressures when the soil is confined. Swelling can arise from the death or removal of vegetation which may be a consequence of climatic change. Although shrinkage increases the intact strength of the clay this is achieved at the expense of a decrease in volume (causing settlement and subsidence) and disruption to mass continuity through cracks and fissures which in turn reduce the mass strength. Vulnerability to swelling and shrinkage can be identified from the Atterberg limit tests (liquid limit and plasticity index). Soils with high liquid limits and high plasticity indices are most susceptible to volume change (ref. 32). For coarser material potential problems such as vulnerability to erosion can be established from the particle size distribution.

Other areas of interpretation requiring attention can be identified from the following section on design.

9. Design and specification
9.1. General
Moving into this phase of the engineering process introduces the need to evaluate all the materials to be incorporated into the works — not just the soils and rock in terms of their suitability for a particular climate and the potential for changes in their properties should the climate change. At the same time there is a need to check that the actions (loads) adopted for design adequately cover foreseeable changes in conditions.

It is easy to view climatic change as being a total world-wide deterioration in conditions. Several texts make the point that this is not so (refs 12, 13). Section 6.3 of this paper examined the regional, localised phenomena, and demonstrated that climatic change would not affect design codes or procedures everywhere to the same degree. A global awareness of existing climate variations and related engineering procedures is important. This is because the influence of climatic change is not expected to one of exceeding phenomena already occurring in the global context. Areas may experience changes in the strength or frequency of strong winds (but probably not their direction in a significant sense). The ability to design for such circumstances probably already exists, therefore, in areas regularly affected by such conditions. Traditional techniques may not suffice in this new climatic regime and it will be these changes, it is suggested, that will cause the greatest difficulty and not the engineering solution to the problem. In this, sense those engineers used to working in an innovative manner are likely to accommodate the effects of climatic change more readily.

9.2. Site preparation
Design work most affected here is likely to be associated with drainage, to accommodate changes in storm frequency and intensity, and the need to provide for access roads not affected by adverse weather conditions. An increase in storminess may also affect temporary works and the sizing of pumps required for dewatering.

9.3. Earthworks and stability
In slope engineering, the ability of clays to sustain negative pore pressures ('suction') due to their low permeability results in a temporary enhancement of stability before the pore pressures return to equilibrium.

Thus the stability of slopes in clay, in contrast to the situation with loads placed on clays, is subject to deterioration with time. The time-frame involved may be significant in the context of climatic change, being some 70 years for failures that have occurred in railway cuttings in London Clay (ref. 16). The significance of this is that the onset of failure is not associated with, and must not be confused with, the climatic changes which may have occurred over this time.

Climatic change may, however, accelerate instability by raising ground-water tables. The significance of the groundwater regime, and surface water, has been examined in section 7 above and design would need to cater for the circumstances applicable.

The design of slope protection and drainage is likely to remain a critical feature of design. Where anchorages are required to enhance stability then special attention would need to be paid to corrosion protection.

Increased storminess or raised temperatures may affect the viability of using fill materials affected by changes in moisture content or specifications would need to be modified to reflect the change in ambient conditions. The addition of strength, using geogrids for example, could find increasing application.

Increased precipitation might prompt a need for the stability of stock-piles and tips to be reassessed. The design of landfill facilities may require reassessment if high temperatures are likely to affect the use or integrity of clay cappings and the raising of water tables may affect containment strategies.

The design of excavations for permanent works could be affected in terms of design water pressures for retaining works and pumping capacity. Cofferdams would need to recognise the potential for higher flood levels or higher tides.

9.4. Foundations

Shallow foundations on soils noted to be susceptible to climatic volume change would need to be founded at depths where moisture content of the soil would not be affected (ref. 34). This is embodied in current good practice but for those buildings with inadequate embedment there is an increasing risk of subsidence with an increasingly dry climate. The processes of swelling and shrinkage are not fully reversible by alternating extremes of temperature or rainfall. Where frozen ground is involved, climatically-raised temperatures would affect shallow foundations and the inundation of dry ground (by rainfall infiltration or by irrigation) would also adversely affect founding conditions (ref. 37).

As with shallow foundations, deep basements and deep foundations (e.g. piles, barettes) could be affected by changes in the ground chemistry brought about by climatic change (ref. 22). Special protection measures would need to be applied in advance of the problem occurring because access to these structural members would not be feasible once the structure had been completed.

The potential problems with deep basements and deep foundations are particularly related to wide-spread changes in the groundwater regime, either due to the ending of water abstraction, when the water table rises (e.g. in London, ref. 27), or the continued and increased abstraction of water, when the water table falls (or, more correctly, when the piezometric profile falls below hydrostatic) (e.g. in Bangkok). With rises in the water table flooding of basements can occur, together with uplift on piles or a deterioration of their load–settlement characteristics. Lowering of the water table or corresponding modification to the piezometric profile generates settlement of the ground and around piles may cause down-drag (negative skin friction).

9.5. Highways

Several aspects of design pertinent to highways have received attention in earlier sections. A lengthening of the 'shut-down' period for earthworks due to inclement weather could see an increased use of capping materials for pavement construction and subsoil drainage capacities would need to be increased. Detention ponds would need to be larger to hold increased runoff, although existing design margins should be examined first. A wider range of temperatures could influence pavement construction (ref. 36) although temperature is likely to affect the design of bridges more if it can be foreseen that the existing design ranges are likely to be exceeded (ref. 33).

9.6. Structures

The most significant design factor affected by climatic change is likely to be wind but direct temperature effects and increased snow loadings would also need to receive attention.

The lightening of superstructures as a result of economies in design has a negative effect on the stability of the building under wind load and increased dependence would be placed on the foundations to resist the potential uplift. In the extreme it may be necessary to supplement the existing foundations with anchorages in order to provide a sufficient margin against overturning. The problem then arises whether or not the anchorages should be pre-tensioned, which places additional permanent axial load on the foundations until such time that the design wind blows, or to leave the anchorage only lightly tensioned and accommodate the deflections necessary before the anchorage force is mobilised during the design wind.

The dynamics of changes in wind patterns — not just extreme values — should also be examined. Effects such as low amplitude 'flutter' and vortex shedding, due to the proximity of structures to each other for example, require attention as this could affect fatigue cycle evaluation and could become more pronounced under climatic change.

Mechanical and electrical provisions, such as additional plant for air conditioning would have an affect on existing structures and future design, as would a lowering in temperatures if cladding insulation requirements were to increase. It should be noted that increased demands for power supply may fight against the attempts to moderate the generation of carbon dioxide, the reason for the climatic change.

10. Construction

Factors influencing construction have received attention as part of the design process described in section 9 above. The effect on construction programmes of stopping work due to inclement weather requires further study (ref. 35). The most significant factor associated with climatic change, in terms of construction procedures, is expected to be precipitation, although markedly stronger winds may necessitate changes to erection techniques for high rise buildings.

Low temperatures can hinder concreting operations and high temperatures may require additives in concrete together with special curing techniques.

The need for monitoring during (and after) construction has been described in 7.6 above and procedures on site would need to accommodate such requirements.

References

1. ANDERSEN A.B. Steel frame and precast floors meet time limit.*Civil Engineering ASCE*, 1976, vol. 46, no. 2, 54–56.
2. *NEW BUILDER*. Row over reduced floor loadings. 1992, 9 July, p. 5.
3. BECKETT D. *Bridges*. Hamlyn, 1969.
4. WHITTEN D.G.A. *The Penguin Dictionary of Geology*. Penguin Books, 1972.
5. GAMMON J.R.A. Weathering of shoreline rock masses — an introduction. *Geological Society of Hong Kong, Bull. 1*, 1984, 35–47.

6. ESCH D.C. and OSTERKEMP T.E. Cold regions engineering: Climatic warming concerns for Alaska. *Journal Cold Regions Engineering, ASCE,* 1990, vol.4, no. 1, March, 6–14.
7. AMERICAN SOCIETY OF CIVIL ENGINEERS. *Civil Engineering* (monthly journal).
8. INSTITUTION OF PROFESSIONAL ENGINEERS, NEW ZEALAND, *New Zealand Engineering Journal* (monthly).
9. WOODCOCK N.H. Geologists and global warming. *Geological Society Geoscientist,* 1991, vol. 1, no.6, 8–11.
10. HOUGHTON J.T., JENKINS G. and EPHRAUMS J.J. *Climate change - The IPCC scientific assessment.* Cambridge University Press, 1990.
11. BARNOLA J.M., RAYNAUD D., KOROTKEVITH Y.S. and LORICH C. Vostok ice core provides 160,000 year record of atmospheric CO_2. *Nature,* vol. 329, 408–431.
12. PETERSON D.F. and KELLER A.A. Effects of climate change on U.S. irrigation. *Journal of Irrigation and Drainage Engineering, ASCE,* 1990, vol. 116, no. 2, March, 194–210.
13. *THE CONSULTING ENGINEER.* Will climate change force new energy strategy? 1981, vol. 45, no.2, February, 9–10.
14. MIMIKOU M., KOWOPOULOS Y., CAVIADAS G., and VAYIANOS N. Regional hydrological effects of climate change.*Journal of Hydrology,* (NL), 1981, vol. 123, 119–146.
15. GRAY V.R. Global warming. *New Zealand Engineering,* 1992, July, 3.
16. VAUGHAN P.R. and WALBANCKE H.J. Pore pressure changes and delayed failure of cutting slopes in over-consolidated clay. *Géotechnique,* 1973, vol. 23, 531–539.
17. TITUS J.G., KUO C.Y. and GIBBS M.J. et al. Greenhouse effect, sea level rise, and coastal drainage systems. *Journal Water Research, Planning, and Management, ASCE,* 1987, vol. 113, no. 2, 216–227.
18. IEA COAL RESEARCH. *Greenhouse Issues.* 1992, no. 3, April.
19. BJERRUM L. *Stability of natural slopes in quick clay.* Norwegian Geotechnical Institute, Oslo, 1955.
20. HAWKINS A.B. *Implications of ground chemistry for construction.* University of Bristol, 1992, July.
21. WEINERT H.H. A climatic index of weathering and its application in road construction. *Géotechnique,* 1974, vol. 24, no. 4, 478–488.
22. BUILDING RESEARCH ESTABLISHMENT. Concrete in sulphate-bearing soils and groundwaters. 1981, *Digest* 250.

23. BROMHEAD E.N. *The stability of slopes*. Surrey University Press, 1986.

24. INSTITUTION OF CIVIL ENGINEERS. *Avoiding disasters*. Hazards Forum, 1992, Third volume, 22 January.

25. PUGH D.T. Is there a sea-level problem? *Proceedings of the Institution of Civil Engineers*, Part 1, 1990, vol. 88, June, 347–366.

26. KOERNER R.M. Remaining technical barriers to obtain general acceptance of geosynthetics (1992 Mercer Lecture). Institution of Civil Engineers, 1992, May (unpublished).

27. CIRIA. *The engineering implications of rising groundwater levels in the deep aquifer beneath London*. Special Publication 69.

28. INSTITUTION OF CIVIL ENGINEERS. *Groundwater Problems in Urban Areas*. 2 to 3 June 1993 (In preparation).

29. HOLMES A. *Principles of Physical Geology*, Thomas Nelson & Son, 1965.

30. THE INSTITUTION OF CIVIL ENGINEERS. *Inadequate site investigation*. Thomas Telford, London, 1991.

31. CLAYTON C.R.I., SIMONS N.E. and MATTHEWS M.C. *Site investigation — a handbook for engineers*. Granada, London, 1982.

32. SIMONS N.E. and MENZIES B.K. *A short course in foundation engineering*. Surrey University Press, 1974.

33. BRITISH STANDARDS INSTITUTION. B5:5400 *Highway Bridges*.

34. BELL F.G.(ed.) *Ground Engineer's Reference Book*, Butterworths, 1987.

35. MORRIS D. Seasonal effects on building construction. *Journal of the Construction Division, ASCE*, 1976, vol. 102, no. 1, March, 29–39.

36. BASMA A.A. and AL-SULEIMAN T.I. Climatic considerations in new AASHTO flexible pavement design. *Journal of the Transportation Engineering Division, ASCE*, 1991, vol. 117, no. 2, 210–223.

37. McKEEN R.G. and JOHNSON L.D. Climate-controlled soil design parameters for mat foundations. *Journal of the Geotechnical Engineering Division, ASCE*, 1990, vol. 116, no. 7, July, 1073–1094.